TURBINE STEAM PATH

MECHANICAL DESIGN AND MANUFACTURE

Volume IIIa

TURBINE STEAM PATH

MECHANICAL DESIGN AND MANUFACTURE

Volume IIIa

William P. Sanders, P. Eng.

PennWell®

1421 S. Sheridan Road
P.O. Box 1260
Tulsa, OK 74101

Managing Editor: Kirk Bjornsgaard
Production Editor: Sue Rhodes Dodd
Cover and Book Design: Robin Remaley

Library of Congress Cataloging-in-Publication Data

Sanders, William P.
Turbine Steam Path Mechanical Design and Manufacture
Volume IIIa / William P. Sanders, P.E.
p. cm.
q.cm
Includes index
ISBN 1-59370-009-1

Printed in the United States of America

1 2 3 4 5 08 07 06 05 04

TABLE OF CONTENTS

Foreword x

Preface xi

Acknowledgments xvii

Author's Note xviii

1 The Basic Considerations of Thermodynamic Design 1

- **Introduction** 1
- **The Thermal Design Process** 3
  - The process of defining the turbine itself 3
  - The basic heat balance 5
  - Other information from the heat balance 22
- **Section and Stage Energy** 26
- **The Design Philosophy and Specification** 28
- **Stage Selection** 31
- **State-Line Efficiency** 38
- **The Velocity (Vector) Diagrams** 42
  - Two-dimensional considerations (pure impulse) 43
  - Two-dimensional considerations (with high reaction)..........49
  - Example 1 (the pure impulse stage) 50
  - Example 2 (a 50% reaction stage) 52
- **Steam Path Sizing and Arrangement** 58

Radial Pressure Gradient 65
Stage Construction Details....................67
Feed Water Heating Trains 69
Flow Splitting and Steam Extraction....................77
Double-flow, high-pressure sections 79
Double-flow control stages in the high-pressure section 80
Extraction of partially expanded steam from the high-pressure section 81
Extraction from high-pressure/reheat double-flow sections 82
Low-pressure extractions from multi-flow sections..............88
Possible low-pressure extraction configurations..................91
References 99

2 The Turbine Steam Path Mechanical Design Process 101
Introduction....................101
The Mechanical Design Process....................102
Information required from the thermal design..................104
Mechanical considerations....................110
Section Arrangement 115
The Major Components of the Steam Path....................118
The Casings 118
The Diaphragms and Stationary Blade Rows 120
Stationary vane design 121
The nozzle plate 122
The Rotors 122
Rotor configuration 123
Technical considerations affecting rotor configuration............126
The Rotating Blades 127
Selection of root form 128

The Stage Hardware 131

The Sealing System 135

Seal system design 135

The Development of New Turbine Products or Processes ..136

Development because of innovation 137

Solving an operational problem 141

The Final Design 143

Stage mechanical analysis 144

Design review 146

Manufacturing specification 147

3 Turbine Steam Path Performance and the Factors Affecting It 149

Introduction 149

Steam Path Performance 151

Steam Path Availability 154

The cost of an outage 158

Interval between inspections 161

State-Line Efficiency 163

Efficiency and the design process 169

Stage leakage losses (Δhl) 171

The effect of leakage steam on state-line efficiency 178

Stage moisture loss (Δhm) 181

Stage leaving loss (Δhv) 184

Basic stage (aerodynamic) efficiency (Δha) 192

Stage Efficiency Considerations 193

Blade Profile Geometry Group One Losses 196

Basic profile loss (ζbl) 196

Stage Reynolds Number (χr) 205

Stage discharge Mach Number (χm, $\Delta\chi m$) 217

Discharge trailing edge (tail)(χb, $\Delta\chi b$) 221

Suction surface trailing edge curvature($\Delta\chi c$, $\Delta\chi s$) 226

Incidence and the inlet nose (χi) 229

Tie or Lacing Wire Losses 235

Stage Geometric Factors Group Two Losses 236

Disc friction (χw) 237

Row aspect ratio (χar) 240

Wall end, flare, and stage transition losses 243

Pressure balance holes 247

Negative root reaction 248

Analysis of the Magnitude of Losses 250

The Cost of Damage Mechanisms 256

Loss due to steam leakage (bypassing of blade elements) 260

Loss due to steam leakage at shaft-end glands 261

Loss due to surface deposits 261

Determination of losses due to profile damage 262

Losses due to discharge area deformation 263

Influence of State-Line Efficiency on Heat Rate 263

References 269

4 Performance Testing 271

Introduction 271

The Purpose of Performance Testing 272

Test Costs and Economics 277

Classes of Performance Test 279

A test requested by the manufacturer 280

Performance test conducted to a specific test code for guarantee purposes 281

Performance test to code standards conducted by the owner....282

A test not conducted to a specific code 283

Test Objectives and Requirements 284

The Enthalpy Drop Test (EDT) 286

Conducting an EDT 286

Using the results of EDTs 290

Other Test Codes 291

The Test Schedule and Advanced Planning 293

The test conference 295

Test Cycle Isolation 300

Test Instrumentation and Calibration 303

Combined uncertainties 303

Measurement of flow quantities 304

Measurement of power output 311

The boiler feed pump power 315

Measurement of thermal energy 316

Measurement of steam quality 319

Measurement of time 320

Measurement of unt speed 320

Simplified ASME PTC-6 Test 321

References 323

Appendix 325

Contents of Volume IIIb 325

FOREWORD

As a companion text to his earlier two-volume set entitled *Turbine Steam Path Maintenance and Repair,* this book represents another valuable contribution to the power generation industry. Engineers involved in all aspects of steam turbine manufacturing, maintenance, and operation will benefit from Mr. Sanders' concise and detailed explanations of thermal and mechanical design principles. In this extensive work, Mr. Sanders covers much of the technical material included in his seminar titled *Turbine Steam Path Engineering.* This popular seminar has been developed over several years and presented to turbine engineers worldwide.

Pressures from industry deregulation and competition have forced many plants to operate beyond their original design life and with fewer resources available for maintenance. Turbine engineers have an increasing range of options available for undertaking steam path repairs, upgrades, or complete replacements. It has become a principle role of the plant turbine engineer to knowledgeably evaluate the options available. In that regard, education through technical resources such as this book will undoubtedly play an important role in this process in upcoming years. Likewise, this text will be an important resource for new steam path designers entering the field because they will benefit from the thorough explanation of all design aspects.

Mr. Sanders' more than 40 years of experience with steam turbines includes design/manufacturing in Europe and the United States as well as utility experience. For the past two decades, Mr. Sanders has been a valuable industry consultant, assisting in projects relating to performance improvement, blade repair options, steam path upgrade strategies, and component failure investigations. Those of us involved in the important field of steam turbine design, performance, and maintenance are indebted to Mr. Sanders for sharing his extensive experience in this text.

Stephen Hesler
Steam Turbine Program Manager
Electric Power Research Institute
Charlotte, North Carolina
April 2003

PREFACE

As energy demands increase, utilities are faced with the responsibility of meeting them. To meet these demands, the providers of this energy are expected to make increased capacity available on their system or through purchases. The general consumers have little or no understanding of what effect increased demands have on those responsible for making power available. The providers of electric energy are not only responsible for making this power available but are also expected to provide this energy at a cost that is acceptable to the consumers. The power suppliers must be prepared to recognize and accommodate this.

The economic health of any community—country, district, or township—is directly related to its access to an adequate supply of electric energy at reasonable costs. The supply must be secure. This energy supply is essential for any industrial or commercial development.

Problems faced by electric utilities in meeting demand are many. First, and most difficult, is that they must be able to anticipate demand, which in a mature community is considerably easier than in a community that is developing and possibly trying to attract different forms of investment. The utility must plan to have adequate energy producing capacity available to support such development. This requirement can become an economic disaster for any community in the event capacity is provided and there is not sufficient demand for that energy.

Another problem is for utilities to determine how best to meet future demand. There are a number of options, each requiring some level of investigation. The alternates for increasing system capacity that need to be considered would include the following.

Installation of new generating capacity

This is ultimately the most expensive option, but it is also the most positive since it will make available on the system equipment that is new, and if selected properly, will comprise the latest technology and would therefore be expected to provide high efficiency and the best reliability available to the users at that time.

However, new capacity cannot be installed in the short term. The time to design, obtain licenses, let bids, and construct the plant will take many years. This is, therefore, the result of long-term planning based on a secure market demand once completed.

Addition of a gas topping unit

There are often a number of units on a system that can be converted to a combined-cycle application. This is an attractive and often cost-effective option for many steam turbine units that have operating life left in them but where other equipment such as the boiler or feed heaters have reached the end of their useful life.

Upgrade an existing base load unit

There are a number of large and mid range units that are currently operating and providing power to the system where unit capacity can be increased. There are two means of achieving this increase, such as redesigning portions of the unit with the installation of major components of improved design that will increase unit output.

It is also possible, if boiler capacity is or can be made available, to increase the amount of thermal energy supplied to the unit by increasing steam flow. This would produce a further increase in output. Unfortunately, such output increases will often result in decreased efficiency because the steam path was not optimized for the increased flow. Some design modifications could be made to improve efficiency; this is a matter of evaluation in terms of the cost of the modifications compared to the amount of increased output. It would also be necessary to evaluate the increase in stress levels of the components installed under the old rating.

Secure a purchase agreement with a neighboring utility

The advantage of this option is that it does not involve any significant and immediate expenditure. However, it is also possible that the costs paid for this energy will be the generating costs of the oldest and most inefficient generators on the system from which the power is being purchased. This could in the long-term result in excessive costs.

This could however, be a suitable solution while a new modern plan is being constructed.

Development of alternate sources of renewable power

The possible development of alternate fuel sources can be considered, such as wind, incineration, hydro, or solar energy. However, several of these concepts are in their infancy of development, and total economics are at this time far from established.

Another problem the utility must address is the need to finance new capacity. This often proves to be the most difficult matter involved in adding capacity. Fortunately, this is beyond the scope of this book, because it would require the wisdom of Solomon to address in any meaningful form. Of the power producing systems available, steam turbines are the most abundant producers in the world today; they are also important motive units in both industrial and marine applications. However, their installation requires the inclusion of considerably more peripheral equipment than the gas turbine or many other possible options. Therefore, their selection may prove to be a more expensive option to providing required power. They can, however, be installed as very large generating units.

A steam turbine generating plant is expensive as a new installation. However, it can be particularly attractive if older units can be upgraded and used in conjunction with a gas turbine in a combined-cycle application.

The development of steam turbines has taken place in a little more than a century. The initial units were built as a consequence of the foresight and pioneer work of engineers such as Parsons, Curtis, and Ljumstrom working independently and each producing a unit that was able to convert thermal potential to rotational kinetic energy. Rotational energy was preferable to reciprocating, which to that time was the most commonly used. Rotational units offered certain advantages. The units are considerably more compact and able to be balanced with a greater accuracy than a reciprocating motion. Although the steam turbine units of these pioneers performed the same function, details of their designs were quite unique. These early efforts laid the groundwork for the preparation of a unit that has now been able to produce outputs up to 900,000 kilowatts (kW) in a tandem compound arrangement and up to 1,300,000 kW in a cross compound configuration.

Many of the units currently in service are contained in power generating plants or industrial installations and continue to operate for years without being shut down for inspection or maintenance. These extended operating periods do not represent negligence on the part of operators. Designers designed units to be opened only every 4 or 5 years for inspection and maintenance. Now, there is a growing tendency within industries to increase this operating period to 8 to 10 years. These extended operating periods are a conscious decision made possible due to advancing technology, particularly in terms of improved materials and the ability of designers to more accurately predict stress levels and the remaining life of components whose failure could force a unit from service.

These extended operating periods have a number of important implications for the operating engineer, including the need to make an exhaustive examination when the unit does become available. To examine these units in greater detail requires they be removed from service and opened for inspection. Reparative actions may have to be undertaken that could have been delayed for a 4-year operating cycle, but cannot for 8 to 10. The cost of opening a unit is high and, more importantly, while these units are open, they are not producing power. Therefore, at those planned outages when units are open it is a responsibility of the plant engineers to ensure all actions identified and required to make the unit suited for another extended operating period are taken.

The extended operating period between maintenance outages provides certain advantages to the owners. They do not require as much reserve capacity, and the cost of opening and taking maintenance activities has in the short-term been reduced. At this time in the evolution of the industry, it is not clear whether, in the long-term, any additional degradation in efficiency during this extended period of operation will cover the cost of the one outage that has been missed. Much of this gain or loss will depend upon the manner in which the plant is operated.

Therefore, the operating staff at any plant is now charged with the additional responsibility of operating the unit to maximize its efficiency and minimize any operating regime that could diminish its reliability or force it from service. Unfortunately, there are limited actions operators can take to achieve these performance goals once the unit has been returned to service. Those positive actions that operators can take include careful monitoring and maintenance of water quality, and helping ensure that the unit does not experience any unjustified or avoidable transients. Many such transients of speed or thermal conditions have the potential to cause deterioration in efficiency and to compromise structural reliability.

Though these difficulties and complexities in generating power exist, the general public has no idea of the difficulties involved in securing an adequate supply of electrical power. Similarly, they cannot comprehend what the possible consequences would be if the engineers responsible for the maintenance and care of power plants were to fail to keep these units running efficiently, reliably, and safely.

There are today many facilities generating with units that were installed more than 40 years ago and that have exceeded their design life. These units are still generating power as required, and this power is fed into the distribution system. Many of these older units are still required to meet peak demands, and this type of duty places a heavy burden on them because the start/stop-type of operation is more destructive as the units are continually cycled. Therefore, while the operating hours may be relatively low, the requirements of maintenance can be high.

In many instances, there may have been significant advances in design methods, manufacturing, and installation since the units being considered were designed. If replacement components are to be purchased, there are often alternate designs or design features available that can either increase efficiency or improve reliability or both. These manufacturing changes can either increase or decrease manufacturing costs. In such instances, it is for the plant engineer to evaluate the advantages of incorporating new components since they always seem to have a higher purchase price. However, the return on investment could well pay for the additional cost within a very short period of operation.

W.P.S.

ACKNOWLEDGMENTS

As in the previous volumes *Turbine Steam Path Maintenance and Repair* I have relied on comments and information from a number of engineers with whom I have worked and cooperated on a number of projects over the past years. Their comments and observations have been of considerable assistance in identifying areas where more detailed information is required and in bringing to my attention subjects that I may not have been, at that time, addressing in sufficient detail, or even at all. For these reasons, their assistance and advice has been invaluable.

As in the earlier PennWell publication, my special thanks go to my friend J. Scott Shearer for allowing me to interrupt his retirement. His review and input from extensive design experience have added considerably to any value other plant engineers may find in this work. His input and the time taken to read and comment on the text and the way it has been arranged has been invaluable.

I would also like to acknowledge the input from the following engineers. Each has made information available to me or has been prepared to discuss certain aspects of units and our joint experiences.

Robert P. (Bob) Dewey and Tony C.T. Lam of Turbine Technologies International; Shannon Ezell of Hickham Industries, Inc.; Thomas J. Hix, Dan J. Stoney, and Cliff. L Baldwin of Omaha Public Power District; Anthony F. Mitola of Turbo Parts, Inc.; Barry McLellan and Blake D. Taylor of SaskPower; Dr. Neville F. Rieger of Stress Technologies, Inc.; and Richard W. Stewart of British Energy.

To make the text clearer, I have used a number of drawings and photographs. Many of these were presented at technical conferences and seminars I attended. For the use of these and other data supplied by manufacturers in their publications I am grateful and have identified these manufacturers where the information has been used.

Special thanks go to the staff members of PennWell for their assistance, especially Kirk Bjornsgaard and Robin Remaley. Their help in particular was invaluable.

W.P.S.

AUTHOR'S NOTE

In 2001/2002 PennWell published a two-volume set entitled Turbine Steam Path Maintenance and Repair. These volumes were based upon a portion of the seminar I present on "Turbine Steam Path Engineering."

This two-volume work covers the remaining portion of that seminar and considers the design options available to the manufacturer and the challenges faced in selecting and arranging the components to optimize unit performance efficiency, reliability, and safety.

The engineer responsible for the health and operation of steam turbines is often faced with the difficult task of selecting between options for repair; these matters are covered in the earlier volumes. This work provides insights into the design options, challenges, and decisions that were originally made by the designers and selection from among these options for the manufacturing facilities at his disposal.

In selecting the information to include, I have tried to anticipate what information a turbine engineer requires when he has to decide what is influencing the performance of his unit or when he is involved with making a decision concerning new components to be incorporated into an existing unit to refurbish or upgrade it for continued use.

Also, I have tried to make these volumes as user-friendly as possible by eliminating as much as possible extensive mathematical treatments. There are a number of excellent volumes that already do this and these are referenced. I have tried to concentrate instead on the practical aspects of the unit—its design and manufacture—to the extent that when problems are encountered or decisions must be made, this reference is available.

For instance, the steam turbine is a thermal machine; it works as a result of expanding high-energy steam and generating rotational energy that is then used to drive some other piece of equipment. For this reason, I have included an introductory chapter in this volume discussing those aspects of thermal design that are of interest to the plant engineer. However, the chapter is not a treatise on thermodynamics.

What I have tried to do is select those thermal considerations that go into designing and arranging the steam path components of the turbine so that it will provide the greatest utility and flexibility to the plant engineer.

There are two departures from this simplified approach, and these are in the areas of steam path efficiency (chapter 3) and rotating blade design (chapter 4). I have elected to provide a more intensive treatment in these areas because rotating blades tend to be of greater concern in the operating unit than any other single group of components and because of the various influences of the geometric stage arrangement and how these differences influence the efficient expansion of the steam.

Turbine manufacturers determine the expansion efficiency of units by establishing the losses that occur as a consequence of stage geometry and steam parameters. Each manufacturer has a methodology for making these assessments. Also, if these methods are compared, there appears to be a degree of variation in them; manufacturers place different weight on the various factors of design, including some and ignoring others. For these reasons, I have provided a greater detail of the factors that influence expansion efficiency. I expect these to be of value primarily to engineers responsible for the thermal performance of the turbine units. I hope this will assist them in determining why any unit appears to be deficient in its performance after some period of operation and enable them to identify some remedial factors that will assist with the restoration of its condition.

Manufacturers' decisions regarding what to include and the weighting of factors is also influenced by field experience and internal research and development. It is surprising but true that if the guaranteed efficiencies of manufacturers are compared, they all tend to finish with a very comparable result.

So, while this work should be considered companion to the earlier PennWell publication *Turbine Steam Path Maintenance and Repair,* certain aspects and subjects covered in the earlier publication are revisited, in outline if not in detail so these two volumes are complete within themselves.

In this volume, chapter 1 discusses those aspects of thermal design that are required to understand component selection and unit arrangement. Chapter 2 gives a brief outline of the mechanical design and development process. Chapter 3 provides information on steam path performance and chapter 4 considers performance testing—not from a how-to-do perspective but rather from a how-to-organize-and-prepare approach.

In volume 2, chapters 5 through 10 consider individual components. Chapter 11 reviews certain aspects of rotating blade manufacture. This is of value because if blades are not manufactured to design specification, they will not operate as intended.

William P. Sanders
Richmond Hill
Ontario, Canada

Chapter 1

The Basic Considerations of Thermodynamic Design

INTRODUCTION

The purpose of this chapter is to outline the thermal considerations responsible for preparing a thermal design specification sufficient that selection and mechanical design can be completed. The object is not, therefore, to prepare a detailed description of the thermal design process. Such description is available in a number of excellent texts, to which the reader should refer. However, it is necessary to provide a brief explanation of some of the more important aspects of design and design choice, so that any differences between the decisions made by turbine builders can be appreciated. It is hoped this allows a more informed evaluation of alternate offerings that are made to the buyer at those times he is evaluating a number of bids and having to choose between them to award contracts.

The steam turbine is a heat engine. It is an engine designed specifically to convert the thermal potential energy of steam to rotational kinetic energy, which can then be utilized to drive a generator or other machine and undertake mechanical work. With current and future anticipated fuel costs, it is essential this energy conversion process is undertaken as efficiently as possible, and for this reason there is considerable pressure applied to current manufacturers to produce units at acceptable costs to undertake this conversion as efficiently and reliably as possible. It can be shown that relatively small changes in element design, arrangement or the incorporation of small changes in cycle detail can, if they introduce an increase in the efficiency of energy conversion represent significant reductions if generating costs. Manufacturers are constantly striving to improve the efficiency of the components of the steam turbine, the cycle, and their arrangement to achieve such improvements.

In this chapter, the thermodynamic principles and basic theory underlying the steam turbine are developed and shown to apply to the rudimentary methods of expanding steam or a two-phase flow of water and steam in a turbine steam path.

The turbine steam path consists of a series of individual expansions or stages each selected to allow a discrete quantity of the total thermal energy to be released and converted to rotational kinetic energy. This energy conversion is achieved in the blade system and is used to drive a mechanical or electric device. There are two distinct concepts of blade system utilized for the release of this energy. These are considered and a rational approach to their design discussed.

The basic design process will examine the individual energy releases, select blade angles and heights for this, and then allow these basic requirements to be examined and refined by the mechanical designer. The designer has a responsibility to ensure the elements are structurally sound in addition to meeting thermal requirements.

In undertaking details of design and making selections of individual components, the design engineer is always faced with constraints. The design process must define a unit that is reliable, safe, efficient, and that can be built and sold profitably at a competitive price level. These goals must be achieved at a cost that allows the manufacturer to compete with other suppliers to the extent he can continue research sufficient to improve the future generations of his units.

THE THERMAL DESIGN PROCESS

The design process begins with agreement of the heat balance, once the supplier and the purchaser have made a final selection of equipment and cycle arrangement. From that time, the detailed design process begins.

The process of defining the turbine itself

A typical thermodynamic design process is shown in Figure 1–1. This shows the design involvement from the preparation of the heat balance to the detailed thermodynamic design. This is a preparation of definition by the thermal engineer, which is sufficient for the mechanical engineer to begin his detailed structural evaluation of the various components and the production of manufacturing details.

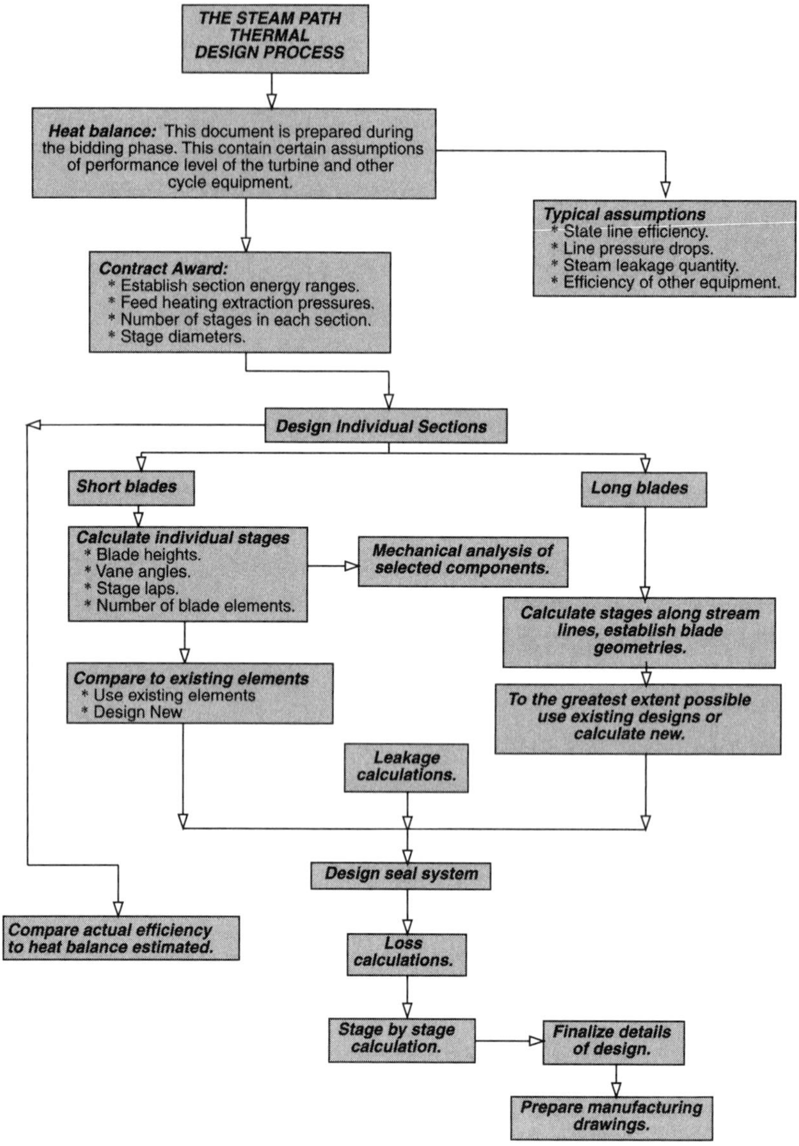

Fig. 1–1 Steam Path Thermal Design Process

The basic heat balance

The power cycle heat balance is a diagrammatic representation of the thermal conditions throughout the power cycle, providing information concerning flow quantities and thermal energy at terminal points. This balance also defines the performance of auxiliary equipment to be used and defines pressure drops and terminal temperature differences where these need to be defined to establish the performance level of the total installation. Therefore, this heat balance provides an account of the energy levels at these various locations. This information is provided in terms of the steam flow quantities and the thermal characteristics (enthalpy, pressure, and temperature). The heat balance also provides information used to size and define the requirements of the various components comprising the cycle, so it is a thermal energy map of the power generating facility.

The heat balance is normally prepared to provide an initial definition of the power cycle. Because the steam turbine is a major component and can often limit certain parameters that are achievable within the plant, it has a considerable influence on the final selection and arrangement of other equipment.

The heat balance is normally prepared during the bidding phase of a contract, which is often before any final selection has been made for the steam turbine and other cycle components to be used. Therefore, at the time in the project of this selection, and before the unit is designed, the turbine supplier must make certain assumptions concerning the arrangement and efficiency of the turbine that will be supplied. In the case of a unit using standard components, this does not represent any major assumptions beyond the general selection and arrangement of the steam path components, the valves and other auxiliary equipment, and steam conditions at terminal points. However, for a new or prototype design, there are a greater number and range of estimates required, all of which can influence the final predicted efficiency and steam conditions throughout the cycle.

In terms of predicting the position and steam conditions at heater extraction points, this is relatively easy in the low-pressure units since the designs are normally modular, with the extraction points selected to provide a suitable thermal gradient throughout the feed heating train. For the high and reheat or intermediate pressure ranges, this becomes a little more complex with estimates being required for the stage points and for steam conditions available for extraction.

Often a heat balance is prepared by the purchaser or his architect engineer and issued with a bid specification. The turbine manufacturer then bids to this heat balance, or, more probably, defines other arrangements and turbine generator configurations using the same performance level auxiliary equipment and main steam parameters identified in the "bid balance," but defining the heat rate for these arrangements. While such proposed cycles may decrease or increase the total cost of the cycle equipment, they also affect the heat rate, and the purchaser would then be expected to evaluate these alternates and determine the most cost effective cycle for his anticipated operating load factor and fuel costs.

The object of this chapter is not to demonstrate how a heat balance can be calculated. Other fully adequate works exist on this subject, including "Effect of Exhaust Pressure on the Economy of Condensing Turbines," by A. Keller and J.E. Downs and "A Method for Predicting the Performance of Steam Turbine Generators 165,000 kW and Larger," by R.C. Spencer, K.C. Cotton, and C.N. Cannon.

Also, with modern computation techniques, such a hand calculation represents a tedious procedure. It is better undertaken by computer methods. These methods are far more flexible and allow the purchaser or owner to make a detailed analysis of the possibility and advantages of changes that could be incorporated. The time spent in making such sensitivity analysis is normally easily justified. This chapter provides some insight into the value of, and the information contained within, the heat balance and how it can be used to interpret the performance of various portions of the cycle and the equipment it contains to assist in making intermediary and operating decisions.

Occasionally, after a contract is awarded, cycle changes or changes in auxiliary equipment that will modify the requirements of the turbine are proposed or requested by either the operator, his architect engineer, or in the case of a prototype turbine design, recommended by the turbine supplier. These changes often result in small effects, both positive and negative in terms of cycle performance, but they should be fully analyzed and the operating costs in terms of both efficiency and the potential effects on availability considered.

The heat balance provided for any installation defines the efficiency of the cycle. This efficiency is a function of three independent factors that interact to define cycle efficiency.

The steam conditions. The steam conditions at inlet to and discharge from the cycle establish the energy available for conversion and help define the power that can be produced from each pound of steam. Because the amount of energy rejected to the condenser is approximately constant for all input energy levels, the ratio to inlet energy conversion increases with increasing inlet energy levels. However, as steam conditions advance, particularly inlet and reheat temperature, there is a greater risk of outage of the unit. The vacuum maintained in the condenser has a significant effect on the cycle performance. (Keller and Downs, 1953)

The efficiency of the equipment comprising the cycle. The individual pieces of equipment comprising the cycle have losses developed in them. These losses are chargeable to the cycle, so as less efficient equipment is used there is less power available at the generator terminals for any stipulated thermal input.

The steam turbine is one of the largest piece of equipment comprising the cycle, (with the boiler and condenser), and as such, its efficiency, and any losses it incurs has a dramatic effect on the total cycle efficiency. Steam path efficiency is considered in greater detail in chapter 3.

The arrangement of this equipment. Some auxiliary cycle equipment drives or is driven either by steam or by electrical power that is chargeable to the turbine. Therefore, it is essential it be located, connected to, and interfaced with the turbine to help ensure its application is optimum and as beneficial as possible to the total generation of cycle output.

A full-load heat balance for an 850,000 kW unit with a double-flow reheat section and a four-flow low-pressure arrangement is shown in Figure 1–2. The heat balance represents the proposed cycle for this installation using initial steam with a pressure of 2415 pounds per square inch absolute (psia), 1000°Fahrenheit (F) initial enthalpy (H) =1460.4, and reheated to 1000°F (H =1519.0). This diagram also shows that the cycle employs seven regenerative feed water heaters, comprising four low-pressure, one direct-contact (deaerator), and two high-pressure elements. This feed water is heated to a final temperature of 483.7°F, using steam taken from the high-pressure section exhaust at a pressure of 587.3 psia. This example unit has steam conditions of 2415 psia/1000°F and an exhaust pressure of 3.65 in. of mercury absolute (Hga).

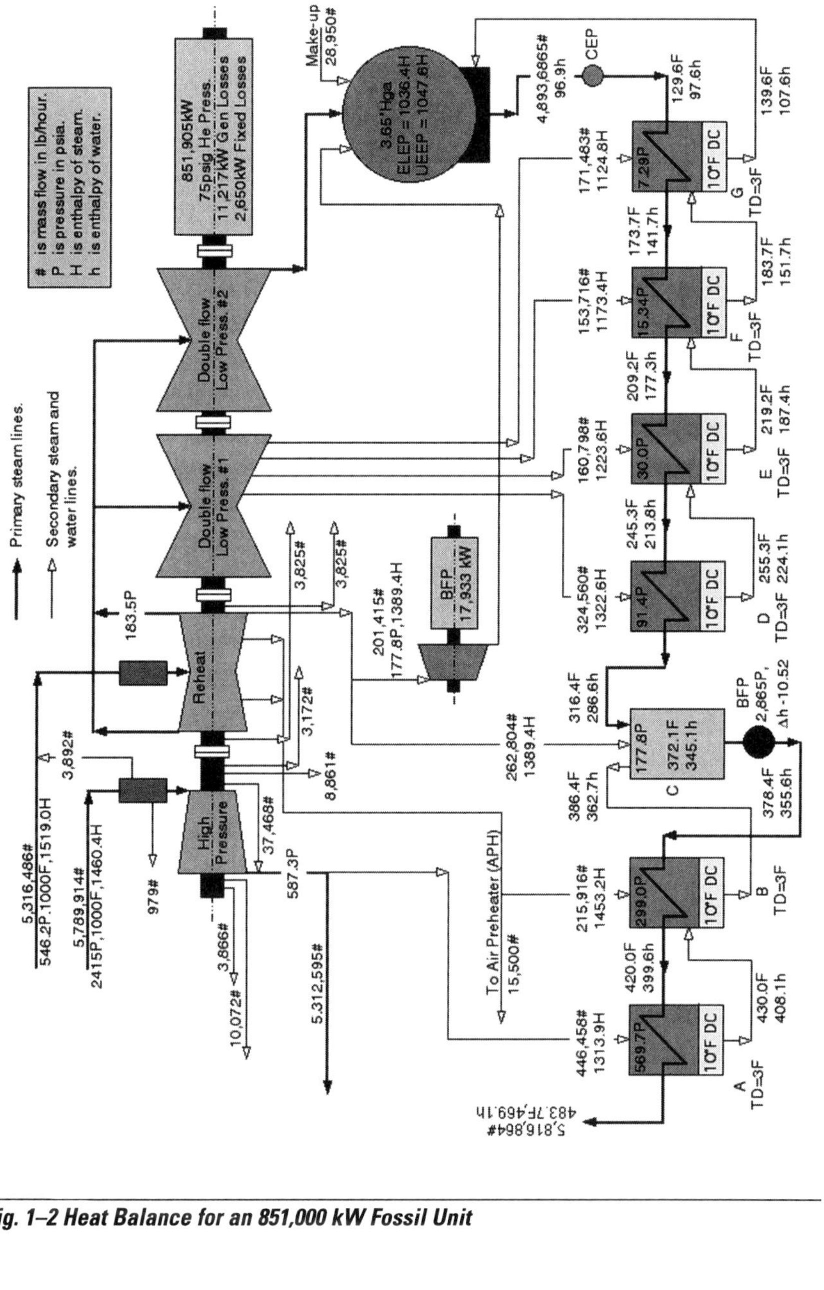

Fig. 1–2 Heat Balance for an 851,000 kW Fossil Unit

The steam passage within this unit, recognizing flow quantities change at various locations within the unit due to extraction and leakage recognize the following:

- Steam is admitted to the high-pressure section with conditions of 2415 psia, 1000°F defining an enthalpy of 1460.4 British thermal units per pound (BTU/lb). The quantity of steam admitted is 5,789,914 (lb/hr.)
- This steam passes through a valve system comprising a stop and control mechanisms. The pressure drop through these values is equal to 3% of the initial pressure. Therefore, the pressure of the steam entering the turbine steam path is 0.97 x 2415 = 2342.6 psia. The enthalpy of this steam remains constant since this loss is a throttling type expansion. There are leakage losses from the valve stems of 979 lb/hr. plus 3892 lb/hr.
- At the high-pressure end of the section, after the steam has expanded through the nozzle plate and entered the rotating blade system, there are steam leakage losses through the shaft sealing system of 37,468, 8861 and 3172 lb/hr.
- The steam expands through the high-pressure section to a pressure of 587.3 psia, with an enthalpy of 1313.9 BTU/lb. At the point where steam is removed from the high-pressure unit, there are steam leakage losses of 10,072 and 3868 lb/hr. through the shaft-end packing.
- At the high-pressure exhaust, steam is removed from the unit and passed to the top high-pressure heater *A*, where the steam is heated to the final feed water temperature (FFWT) of 483.7°F, having an enthalpy of 469.1 BTU/lb.
- Upon removal from the high-pressure section, the steam is returned to the boiler reheater section where its temperature is increased to 1000°F. The steam is then returned to the

reheat section of the turbine with an enthalpy of 1519.0 BTU/lb. In passing through the reheat portion of the system, comprising the cold and hot reheat lines and the reheater portion of the boiler, there is a 10% drop in the steam pressure. Also the quantity of steam flowing to the turbine has been reduced to 5,316,486 lb/hr. Also flowing into the reheat turbine, there is a quantity of steam taken from the control valve leak off that is 3892 lb/hr.)

- The steam entering the reheat section divides into two parallel steam paths and expands to an exhaust condition of 183.5 psia at an enthalpy of 1389.4 BTU/lb. In this section, after partial expansion, steam is removed from both the turbine and generator end flows to supply steam to the heater indicated as *B*. Also steam is taken from the exhaust to supply feed heating steam to heater *C* (deaerator). Steam from this exhaust, seen in our example at 201,415 lb/hr., is also used to drive the boiler feed pump turbine (BFPT). After expanding through the BFPT, the steam is exhausted into the main condenser.
- At both shaft-end positions, there are leakage losses of 3825 lb/hr.
- The steam removed from the reheat section of the turbine is then passed into a crossover pipe where it is delivered to the low-pressure section. This steam divides into four portions, and expands through the low-pressure blading. There are four extractions from the low-pressure sections passing steam to heaters *D*, *E*, *F*, and *G*.
- The steam exhausts from the low-pressure section through blades that have an active length of 31.5 in. The expansion-line end point is 1036.4 BTU/lb., and the used-energy end point is 1047.6 BTU/lb. The quantity of steam flowing to the condenser from the low-pressure sections is 3,821,580 lb/hr.

The expansion line end point (ELEP) defines the enthalpy of the steam at exhaust from the low-pressure turbine sections. However, in the exhaust from the low-pressure section, energy is lost due to the velocity of the steam entering the condenser. This lost energy is deducted from the total exhaust energy, and the ELEP defines the used energy end point (UEEP) as seen in chapter 3.

From the heat balance, other information can also be determined, relating to both the output of the various sections and the cycle configuration. As an example of the output determination, consider the high-pressure section:

- The basic high-pressure section is shown in Figure 1–3, and the expansion line for this same section in Figure 1–4. This expansion line shows the effect of the 3% pressure drop at constant enthalpy through the inlet valve system. This pressure drop increases the steam entropy from 1.5324 to 1.5352 ft-lb/lb/°F. Associated with this pressure drop there is also a reduction in steam temperature from 1000 to 996°F.

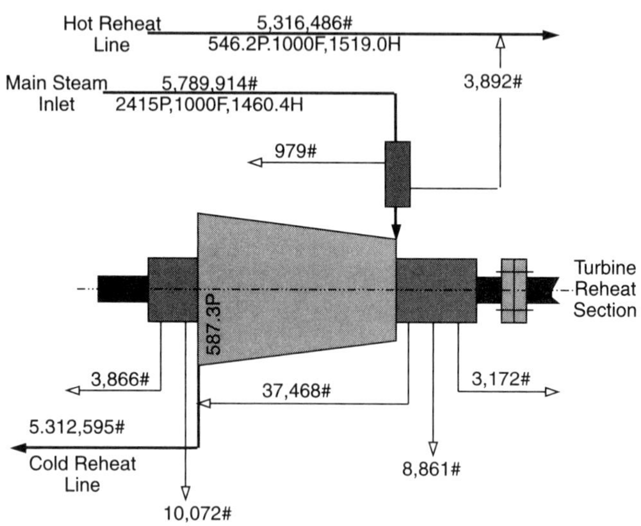

Fig. 1–3 Details of the High-Pressure Section for the Unit Shown

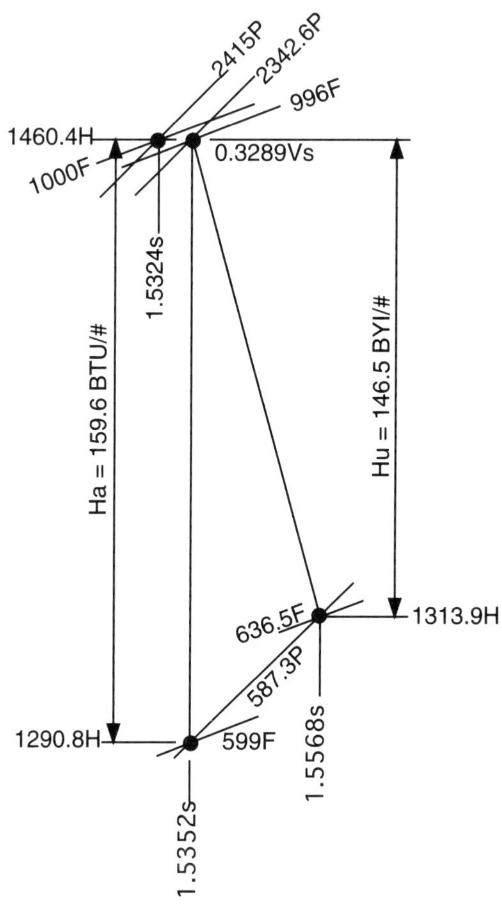

Fig. 1–4 The High-Pressure Section Expansion Line

- The steam expands in the high-pressure steam path to a pressure of 587.3 psia, having a temperature of 636.5°F and an entropy of 1.5568 ft-lb/lb/°F. The isentropic enthalpy drop from a pressure of 2342.6 to 587.3 psia is 169.6 BTU/lb.

Therefore, the expansion line efficiency ηsl is:

$$\eta sl = \frac{\text{Useful Enthalpy}}{\text{Available Enthalpy}} \tag{1.1}$$

$$= \frac{1460.4 - 1313.9}{1460.4 - 1290.8} = \frac{146.5}{169.6} = 86.4\%$$

Also from this expansion line data, the following physical characteristics of the steam and its expansion can be determined:

section pressure ratio: = 2342.6/587.3 = 3.989

specific volume at section inlet = 0.3289 cu ft/lb

specific volume at section discharge: = 1.0161 cu ft/lb

The approximate output of the high-pressure section can also be determined from the steam properties and the steam mass flow. The enthalpy which is useful is equal to 1460.4–1313.9 = 146.5 BTU/lb.

The mean weight flow, *mhp* in the high-pressure section, is the mean flow at inlet, which is equal to the flow at inlet of the steam delivered from the boiler minus the valve steam leakage quantity. The steam flow at discharge is equal to flow at inlet minus the seal steam leakage at the high-pressure end.

steam flow at inlet = 5,789,914 - (979 + 3892) = 5,785,043 lb/hr.

steam flow at discharge = 5,785,043 - (37,468 + 8,861 + 3,172) = 5,735,542 lb/hr.

$$mhp = \frac{5{,}785{,}043 + 5{,}735{,}542}{2} = 5{,}760{,}293 \text{ lb/hr.}$$

In fact this flow of 5,785,043 lb/hr determined for the flow at inlet is not absolutely correct, because at discharge from the control stage there is a shaft-end leakage of 37,468 + 8,861 lb/hr, and the flow calculated and used as the inlet flow should be corrected for this gland leakage at the high-pressure end. (That is, this calculated quantity of 5,785,043 lb/hr goes through the first or control stage only.)

If the additional flow through the control stage is neglected, then the inlet flow is the same as the discharge flow at 5,735,542 lb/hr as there are no internal extractions. Therefore, the output can be calculated using both inlet flows as:

$$kW1 = \frac{5{,}760{,}293 \times 146.5}{3412.14} = 247{,}317.8 \text{ kW}$$

$$kW2 = \frac{5{,}735{,}542 \times 146.5}{3412.14} = 246{,}255.1 \text{ kW}$$

Note that 3412.14 is the number of BTU/hr units in 1 kW.

In fact, the actual output is somewhere between these two values and also modified by internal leakage under stationary blade rows and over the rotating blade tips. These quantities cannot be determined from the heat balance, but the previous figures provide some indication of the high-pressure section output.

Similarly, output for the reheat and low-pressure sections can be determined. However, in these sections the quantity of steam flowing is not constant throughout the expansion, and it becomes necessary to take the extraction quantities into account.

Consider the schematic of the reheat section shown as Figure 1–5, which includes extraction quantities and steam conditions around heaters *B* and *C*.

There are two steam extractions, the first at a pressure of 318.1 psia, and the second at exhaust from the section.

The steam flowing into the section is: = 5,316,486 + 3,892

= 5,320,378 lb

The enthalpy drop on this first section is: = 1519.0 - 1453.2

= 65.8 BTU/lb

The output of this first section is: $= \dfrac{5{,}320{,}378 \times 65.8}{3412.14} = 102{,}598.6$ kW

For the second section, between pressures 318.1 psia, (1453.2 BTU/lb) and 183.3psia, (1389.4 BTU/lb), then:

The steam flowing through it is: = 5,320,378 - 215,916

= 5,104,462 lb

The enthalpy drop on this second section: = 1453.2 - 1389.4

= 63.8 BTU/lb

The output of this first section is: $= \dfrac{5{,}104{,}462 \times 63.8}{3412.14} = 95{,}442.9$ kW

Therefore, the total output of the reheat section = 102598.6 + 95442.9

= 198041.5 kW

A similar analysis can be made for the low-pressure section with its four extractions, which may not be symmetrical and therefore more complex.

There are data available from the heat balance that do not apply to the steam turbine and generator but do influence the heat rate and the

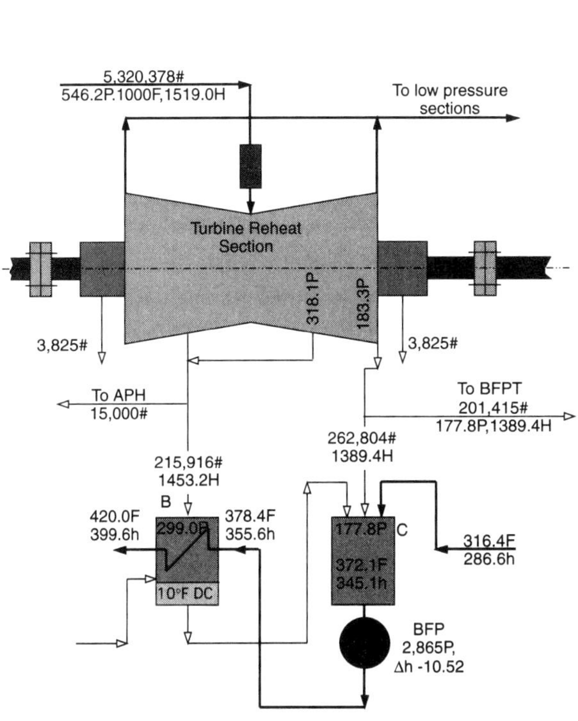

Fig. 1–5 The Double Flow Reheat Section

efficiency of the power cycle. Typical of these other characteristics is the feed heating cycle and the performance of the individual heaters. Typical of this information on regeneration within the cycle are:

- The condensed steam is removed from the condenser by a condensate extraction pump (CEP), and pumped through the low-pressure heaters. This pump develops sufficient head to deliver the condensate to the deaerator, which is located at an elevated location within the plant. Upon removal from the deaerator, this water is pressurized by the 17,933 kW boiler feed pump (BFP) to 2865 psia. This BFP is shown in Figure 1–2. The pump is driven by a BFP turbine that utilizes 201,415 lb of steam extracted from the reheat section exhaust. Upon completing its expansion, this steam is exhausted and returned to the main condenser.

- The heater train for this cycle is shown in Figure 1–6. Portion (a) shows the heaters with flows and steam conditions, and (b) shows the thermal rises through the train. In Figure 1–6 is shown the heaters comprising the cycle and shown in Figure 1–1 with flows and temperatures at various locations. These heaters are arranged so the low-pressure elements have a terminal temperature difference (TTD) of 3°F and a drain-cooled section with a temperature difference of 10°F. Similarly, the high-pressure heaters have a TTD of -3°F and a similar arrangement in the drain section.
- There is a pressure drop in the lines connecting the heaters to the turbine extraction points. Consider the heater using steam extracted from the reheat section at 318.1 psia. At the high-pressure heater, the inlet pressure is 299.0 psia, indicating a 6% pressure drop. Similarly, the steam extracted at 97.2 psia from the low-pressure section has a pressure at the heater of 91.4 psia, again indicating a 6% pressure drop. Different pressure drops could have been used. But this affects pipe size and therefore plant costs. It is always necessary for the engineer responsible for defining plant parameters to optimize these various costs against unit performance.

Figure 1–6 shows the thermal gradient throughout the feed heating train, plotting the temperature and enthalpy rise. These values are also shown in Table 1–1.

Heater:	G	F	E	D	C	BFP	B	A
Temperature at Inlet:	129.6	173.7	209.2	245.3	316.4	372.1	378.4	420.0
Temperature at Discharge:	173.7	209.2	245.3	316.4	372.1	378.4	420.0	483.7
ΔT through equipment:	44.1	35.5	36.1	71.1	55.7	6.3	41.6	63.7
Enthalpy at Inlet:	97.6	141.7	177.3	213.8	286.6	345.1	355.6	399.6
Enthalpy at Discharge:	141.7	177.3	213.8	286.6	345.1	355.6	399.6	469.1
ΔH through equipment:	44.1	35.6	36.5	72.8	58.5	10.5	44.0	69.5

Table 1–1 Thermal Increases in the Heater Train

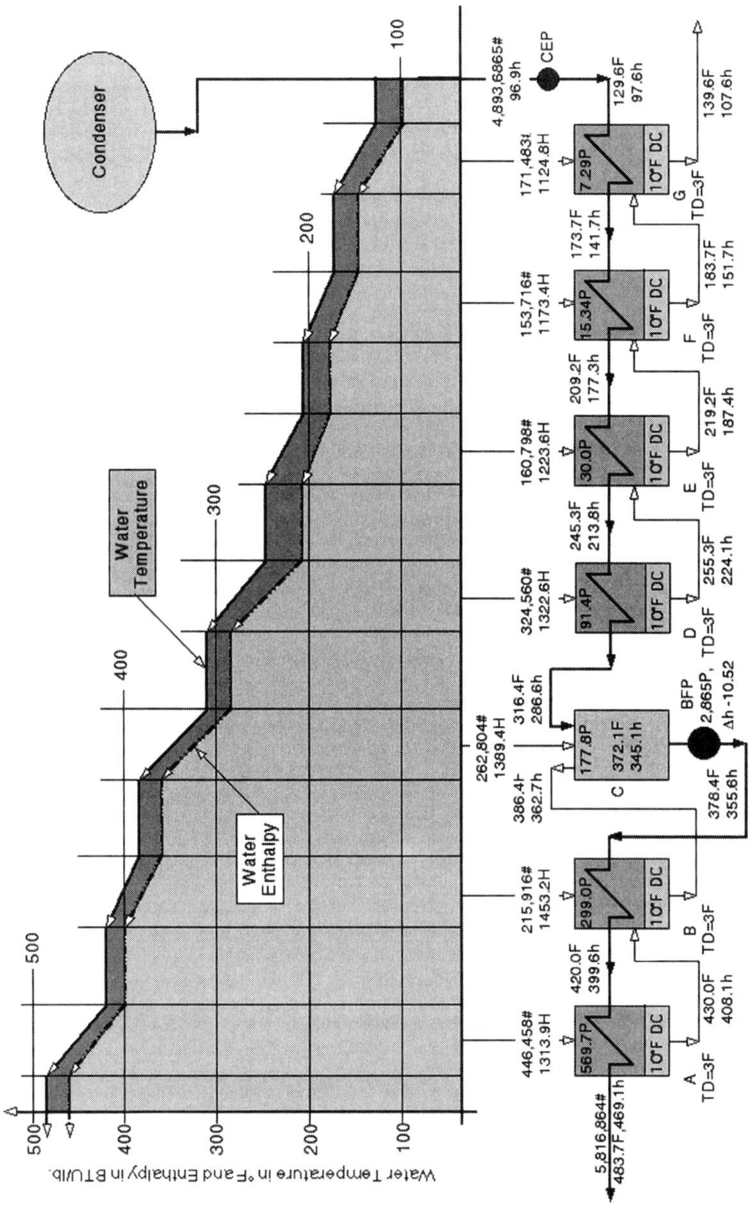

Fig. 1–6 The Heater Train for the Seven-Heater Cycle

- The individual feed heaters are heat exchange vessels, and it is possible to make a heat balance around each of them. Consider the top heater generating the FFWT, utilizing steam extracted from the high-pressure section exhaust. The thermal conditions around this heater are shown in Figure 1–7. This figure shows the top heater of the feet train from Figure 1–2. The FFWT is 483.7°F. For this, like other heaters, there is a thermal balance between the energy transferred from the heating steam extracted from the turbine and the feed water being returned to the boiler.

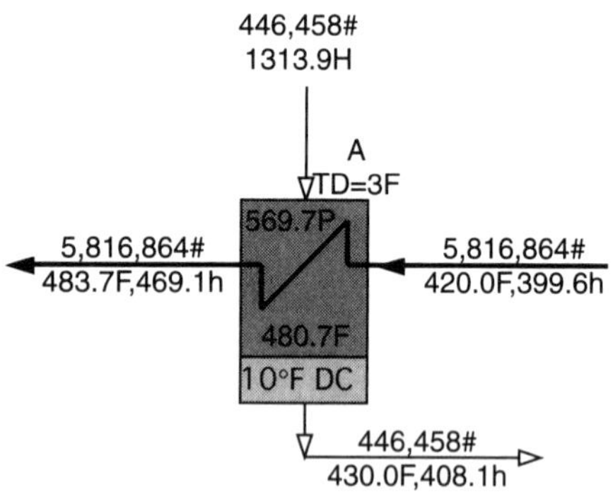

Fig. 1–7 The Top Heater of the Feed Train

There is also other information supplied on the heat balance as shown in Figure 1–8. Here is shown the steam flows to the steam seal regulator (SSR) and the steam packing exhauster (SPE). From this portion of the balance it can be seen that four different leakages from the shaft sealing point and one from the control valves are piped to the SSR. This steam is used to seal the system at start-up, and provide sealing steam to the low-pressure section glands at all loads. Regarding this seal system, the following points can be noted:

- The SSR receives a total flow of 3825 + 3825 + 3886 + 979 + 3172 = 15,687 lb of this flow, and, at this load condition, 2400 lb is passed directly to the condenser while 2800 lb is passed through the SPE, where it is used to increase the energy level of the feed water leaving the condenser, raising its temperature from 129°F to 129.6°F. The remaining 10,487 lb is sent to the lowest pressure heater where it is used to preheat the feed water.
- Shown in Figure 1–2 is a make-up quantity, which is defined as 28,950 lb/hr in Figure 1–8. This quantity is to replace any losses of water or steam from the cycle, which is assumed to be 0.5% of the steam inlet quantity. That is 0.005 x 5,789,914 = 28,950lb/hr.

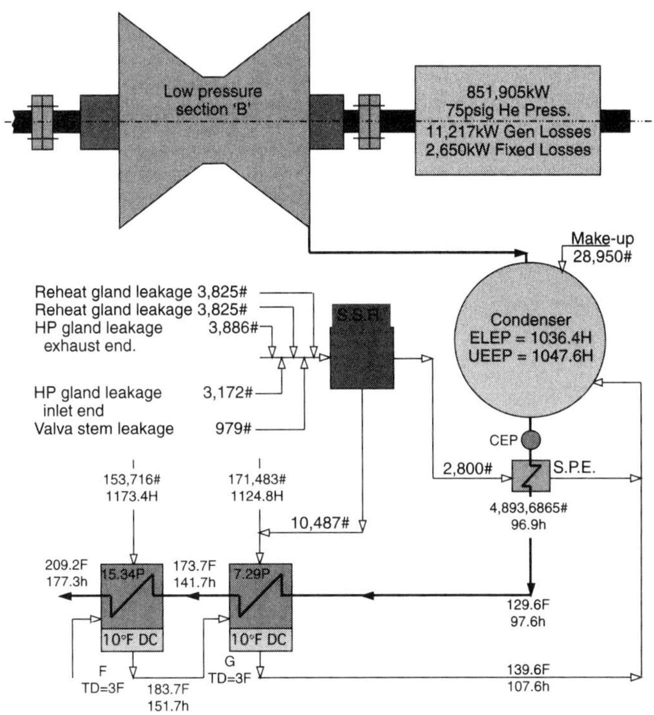

Fig. 1–8 The Steam Seal System Showing Flows and Thermal Conditions

A method for predicting the performance of units rated 165,000 kW and larger was published, and has now been computerized to allow quick determination of the information required on any heat balance. (Keller and Downs 1953) (Spencer, Cotton, and Cannon 1962)

Other information from the heat balance

In addition to information on section efficiency and output, other important information can be determined from the heat balance, and it has the capability of allowing the plant engineer to evaluate the performance and losses, which have a direct effect on the overall performance of the turbine generator.

The valve-stem leakage. In the heat balance Figure 1–2 it is shown that the main stop and control valves have two leakages, a high-pressure leakage of 3892 lb/hr that is returned to the hot reheat line to enter the reheat section of the unit, and a leakage at a lower pressure of 979 lb/hr that is led to a lower energy level, often the steam-seal regulator.

Consider the first leakage of 3892lb/hr. This steam bypasses the entire high-pressure section and therefore does no work in this section. That is, it bypasses an enthalpy drop of 1460.4-1313.9 = 146.5 BTU/lb. This represents an output reduction *kWvi* of:

$$kWvi = \frac{3,892 \times 146.5}{3,412.14} = 167.1 \text{ kW}$$

This is a normal leakage amount and is allowed for in the basic design rating of the unit when the valves are assembled with normal clearances between valve stems and bushings. However, if these clearances increase due to wear, it is obvious that relatively small increases can cause a significant decrease in output of the high-pressure section.

Similarly the lower-pressure leakage of 979 lb/hr bypasses the entire expansion in the turbine from 1460.4-1036.4 = 424 BTU/lb. This represents an output loss of:

$$kWvi = \frac{979 \times 424.0.5}{3{,}412.14} = 121.7\ kW$$

From these numbers it can be seen that valve maintenance has a significant effect on unit output.

Shaft-end packing leakage. At the shaft-end position of the high-pressure inlet end, there is a total leakage along the shaft of:

$$37{,}468 + 8861 + 3172 = 49{,}501\ lb/hr$$

Again this steam bypasses the high-pressure section causing an output loss *kWL* loss of:

$$kWL = \frac{49{,}501 \times 146.5}{3{,}412.14} = 2{,}125.3\ kW$$

This same analysis is undertaken for each shaft-end position in the turbine train. However, again this loss is anticipated and allowed for. The unknown are those losses that occur as a consequence of clearance increases at the shaft end position.

Combined rotor leakage. In many designs rotor portions are contained on a common shaft. Such a configuration is shown in Figure 1–9, where the rotor is a combined high and reheat section. In this design, the steam for both expansions is admitted to the unit at the center, and since the design of the intermediate pressure section is reheat, then the temperature at the center section is common (within small levels of difference). However, the pressure at this center location is different, so it becomes necessary to provide a sealing arrangement at this location to limit the amount of steam leaking along the shaft from the high to reheat sections. In the figure this is shown as the quantity *m2*. There are also the shaft-end leakage quantities *m1* and *m3* at the exhaust ends from both the high and reheat sections.

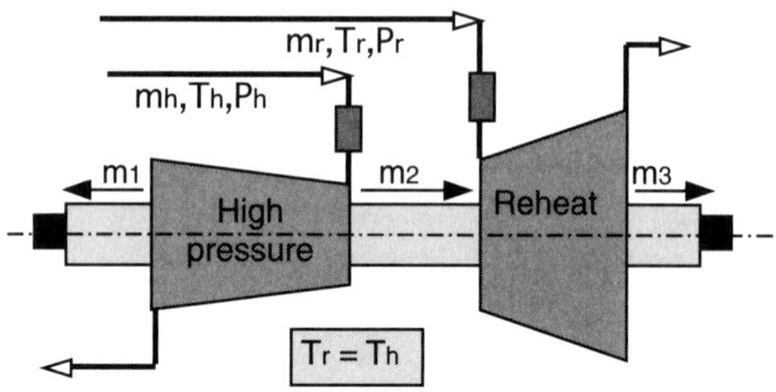

Fig. 1–9 The Section Arrangement of a Combined HP/Reheat Rotor

At inlet to the high-pressure section the steam quantity flowing is *Mh*, at conditions *Th* and *Ph*. The steam leaves the high-pressure section and is returned to the boiler reheater section and then returned to the turbine at steam conditions *Tr* and *Pr*. At this center section, therefore, there is an approximate pressure difference of $\Delta P = Ph - Pr$. This value is not absolute because the pressure at the high-pressure inlet has been reduced due to the pressure drop through the control stage nozzles, but the pressure drop through the reheat section is accounted for by establishing the value of *Pr*. There will, however, be a small drop in pressure through the intermediate stop and control valves.

Therefore, at the center section, this pressure differential will drive steam through the gland system from the higher- to the lower-pressure level. It is interesting to consider the effect this leakage quantity, shown as *m2* in Figure 1–9, will have on the unit performance.

Consider the expansion line shown as Figure 1–10. The steam conditions of Figure 1–9 are represented on this diagram, and the pressure differential ΔP is shown. In this figure, there are a number of pressure drops. These are

δPh	the pressure drop in the main valves
Ph - Phc	the pressure drop through the control stage nozzle
δPbr	the pressure drop through the reheat system
δPr	the pressure drop through the intermediate valves
Pr - Prc	the pressure drop through the first stage of the reheat section
ΔP	the pressure drop sensed by leaking steam *m1*

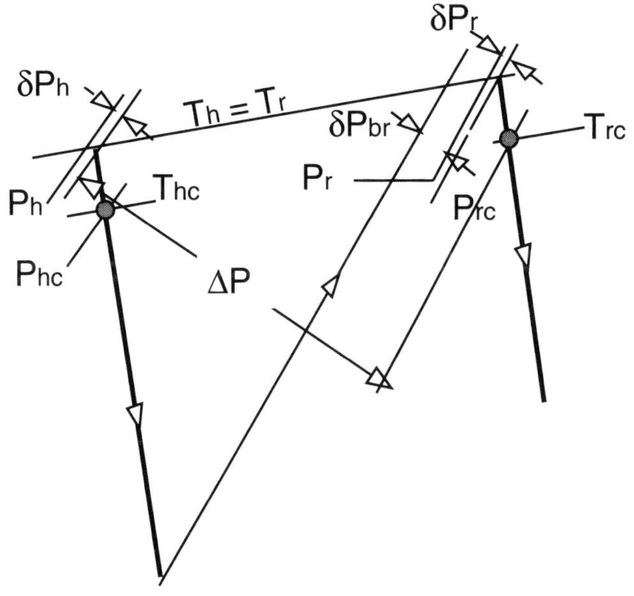

Fig. 1–10 The Mollier Diagram for the Turbine Sections

From this expansion diagram, it can be seen that the steam that leaks across the seals carries with it energy that degrades the output of the high-pressure section, bypassing the entire rotating blade rows, but that can be utilized to produce power in the reheat section.

The net effect of this leakage is to degrade the high-pressure section efficiency and output and to increase the output and efficiency of the reheat section. However, the overall effect on the unit is a degradation of output, since the leakage quantity bypasses the high-pressure section, and while it does generate output in the reheat section it would have passed through this section of the turbine and produced the same level of power anyway.

SECTION AND STAGE ENERGY

Before beginning the detailed design process for the individual stages, it is necessary for the design engineer to first establish the energy ranges of the individual sections and then the details of the stages. There are various considerations related to the selection of the high-pressure extraction pressure, including the possible requirements of removing steam at a pressure that will provide heating steam to achieve the final temperature of the feed water.

It is necessary to consider the general process of selecting the energy ranges of the various turbine sections and where steam should be removed from the unit and returned to the boiler for reheating. At this point in the design, no effort has been made to define the optimum arrangement of the stages, number of stages, or diameters. These requirements are considered in the next section.

However, from the Mollier diagram in Figure 1–11, it can be seen that assuming the high-pressure extraction pressure has no impact upon state-line efficiency, then the extraction pressure will have an impact on the total unit design. Note the following specific effects.

- As the extraction is lowered, the final moisture content in the low-pressure section exhaust is increased. It thus has the potential to increase the moisture damage in the back end of the low-pressure sections.
- As the high-pressure section extraction is lowered, there will be a requirement for larger piping from the high-pressure exhaust to the boiler reheater and from the reheater back to the reheat section of the turbine, because of the steam's larger specific volume. This will also increase the size of the reheater tubes required in the boiler reheater. If this piping is not increased in size, the steam velocity in these pipes will increase, boosting the frictional loss in the lines because it is dependent upon velocity squared.
- This also has the potential to influence the axial thrust in the various sections and the size of the thrust block required.

The effect of the high-pressure expansion from pressure *Pi* to *Pr* is seen in the figure with five alternate extraction pressures, seen here as *Pe*. Their effect on steam conditions throughout the steam path can be seen down to the condenser pressure *Pc*, with moisture contents *X1* through *X5*.

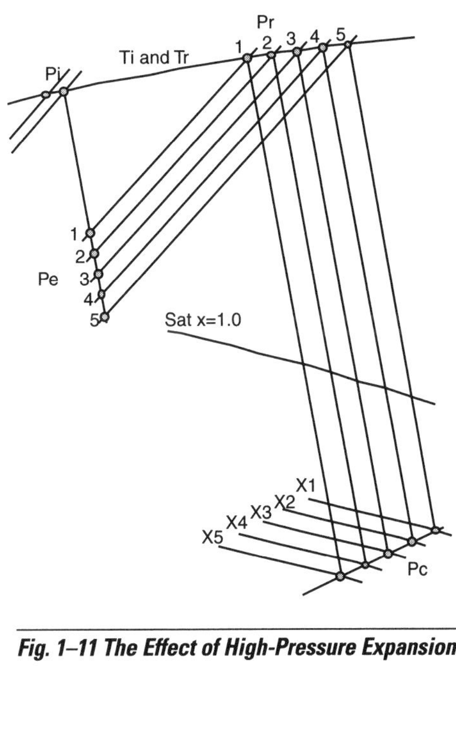

Fig. 1–11 The Effect of High-Pressure Expansion

DESIGN PHILOSOPHY AND SPECIFICATION

There are two distinct philosophies of design. These philosophies are based on the manner in which thermal energy is released in the stages. These two philosophies are termed *Impulse* and *Reaction*. In the pure impulse stage, the entire thermal energy is converted to a kinetic effect in the stationary blade row, and the rotating row simply converts this kinetic energy to thrust, primarily in the tangential direction, by turning the steam through as large an angle as possible and driving the rotor. In the reaction stage, a portion of the thermal energy is converted to kinetic in the stationary row, which helps drive the rotor with the remaining portion released in the rotating row, thereby producing a reaction on the vane.

In fact, the steam paths of these two units normally utilize both philosophies to a degree to allow overall design requirements to be met. The reaction units will normally employ an impulse design in their first high-pressure, or control, stage. This is done, among other considerations, to allow a larger pressure drop through the first row of stationary blades and thus subject the casing to lower pressures and temperatures. In the impulse design, it is common to allow some small degree of reaction at all radial heights of the stage so there is always a positive pressure difference across each blade row down through the stages.

These differences in design concept result in blade vanes and entire steam path construction that are quite different. Consider the factors contributing to these differences as follows.

Energy release in the stage. The optimum energy release in a stage to achieve a maximum efficiency is dependent upon a number of factors, most significantly on the ratio of blade tangential velocity to steam isentropic velocity equivalent. This ratio is normally given the symbol ρ. The efficiency is also influenced by secondary factors such as the steam discharge angle $\alpha 1$, from the stationary blade row, and the flow coefficients.

Equations have been developed to show that for a 50% reaction design that the gross stage efficiency is achieved when:

$$\eta_{gs} = \frac{\eta s . \left[2\rho . \mathrm{Cos}\ \alpha 1 - \rho^2 \right]}{\left[1 - ks^2 \left[1 + \rho^2 - 2\rho \mathrm{Cos}\ \alpha 1 \right] \right]} \tag{1.2}$$

This stage efficiency is maximized when the value of ρ is:

$$\rho_{opt} = \mathrm{Cos}\ \alpha 1 \tag{1.3}$$

where

ρ = U/Co

U = blade tangential velocity

Co = the velocity equivalent of the stage isentropic drop

Similarly, for the pure impulse stage, the optimum value of efficiency is:

$$\eta_b = 2\left[1 + \Phi v \right] . \left[\rho\ \mathrm{Cos}\ \alpha 1 - \rho^2 \right] \tag{1.4}$$

where

Φv = frictional loss coefficient

In equation 1.4, this does not consider the losses that would occur in the stationary blade row.

However, this stage efficiency is maximized when the value of ρ is:

$$\rho_{opt} = \frac{\text{Cos}\ \alpha 1}{2} \tag{1.5}$$

where

$\alpha 1$ = stationary blade row discharge angle

Φv = row flow coefficient

To establish what these differences in optimum ratio of ρ do to stage arrangement, consider two stages both with the same flow and velocity coefficients. A comparison of ρ is shown in Figure 1–12. These same curves are also shown on Figure 1–13 for the reaction stage and on Figure 1–14 for the impulse.

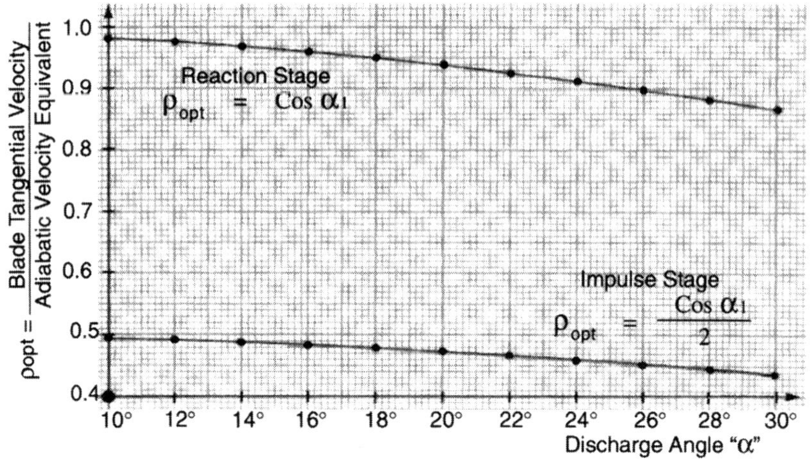

Fig. 1–12 The Optimum ρ Values for a Reaction and Impulse Stage

From this figure, and from comparisons of equations 1.3 and 1.5, it can be seen that the optimum ration of ρ is twice as high in the 50% reaction stage as in the pure impulse. To examine the significance of this, consider that a turbine section operating at 3600 revolutions per minute (rpm) is to have an energy range of 163 Btu/lb, and the stages are to have a mean diameter of 36 in. A comparison of stage selection is shown in Table 1–2.

To make this comparison, it is necessary to have available certain conversion factors.

These are:

Rotating Blade Tangential Velocity *U.*

$$U = \frac{2 \times \pi \times 60}{3600} \cdot \frac{Dm}{2 \times 12} = 15.71 \, . \, Dm \qquad (1.6)$$

where

Dm = the stage mean diameter, in inches

Steam Isentropic Velocity *Co.*

$$Co = \sqrt{2.g.J.\Delta h} = 223.7\sqrt{\Delta h} \qquad (1.7)$$

where

g = gravitational acceleration constant

J = Joules mechanical equivalent of work

ΔH = stage enthalpy drop

Stage selection

From equations 1.6 and 1.7, it can be seen that the total energy on a section should be divided between the stages so as to maximize the efficiency of the section, which requires that the value of ρ is optimized. Consider the situation in Table 1–2.

	Reaction	Impulse
Stage diameter	36.0"	36.0"
Tangential velocity, ft./sec.	565.6	565.6
Steam discharge angle $\alpha 1$	16°	16°
U/Co = ρ	0.961	0.481
Stage isentropic velocity, ft./sec.	900.7	1175.9
Stage enthalpy, Btu/lb	16.21	27.63
Number of stages = 163 Btu/lb	10.06	5.90

Table 1–2 Comparison of Reaction and Impulse Stages

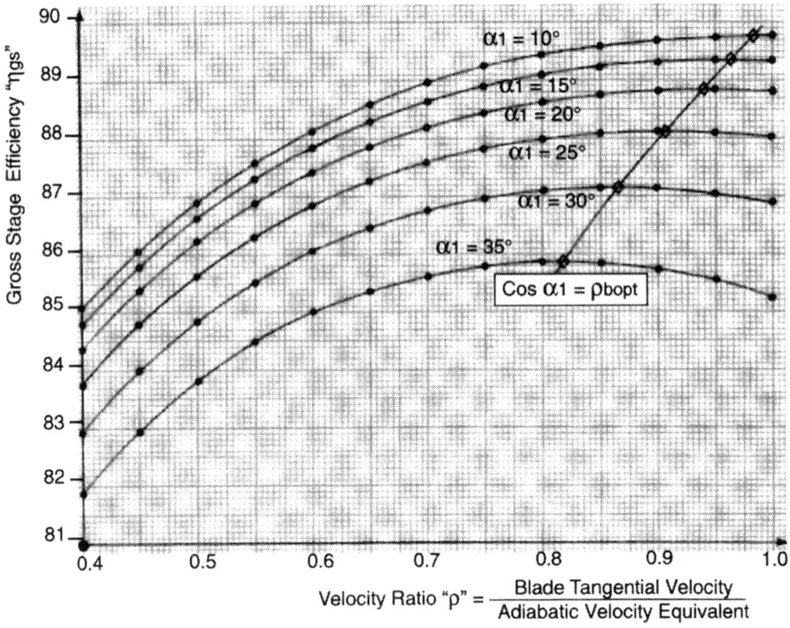

Fig. 1–13 Efficiency of a 50% Reaction Stage as a Function of the ρ Ratio for Different Steam Discharge Angles

From this comparison, it is clear that the 50% reaction design requires more stages than the pure impulse. However, in a practical design, partial stages are not possible, so some small amount of compromise must be made in mean diameters. But it is unlikely that any design would be produced with each stage having the same mean diameter. Another consideration is that the pure impulse stage is very rare in large units so the enthalpy per stage will be larger, making a requirement for a larger number of stages. Also, had the energy range being considered been a high-pressure section with a control stage, then a larger portion of the total enthalpy would have been expended across that stage, reducing the number of reaction stages required. Figure 1–13 displays the efficiency of a 50% reaction stage as a function of the ρ ratio for different steam discharge angles.

Sensitivity of stage efficiency to ρ. From the expressions for efficiency of the stage in the case of the reaction design and for the rotating blade in the impulse, it is possible to review the impact of the value of the ρ ratio on the stage and how any modification will influence performance.

- **The reaction design.** Using equation 1.2 curves, Figure 1–13 is constructed showing the variation of gross stage efficiency ηgs for various steam discharge angles $\alpha 1$. Also shown on this curve is the locus of maximum efficiency for each discharge angle from 10° to 35°. The effect of the discharge angle on efficiency can be seen. This magnitude of change provides an incentive for the designer to make the discharge angle as small as possible, consistent with being able to contain the blade radial height. Any reduction in discharge angle will close the throat and reduce discharge area. This throat reduction must be compensated for by an increase in vane radial discharge height.

- **The impulse design.** The impulse stage is examined using equation 1.4 and plotting this function as Figure 1–14. Here we use the same stage parameters as were used for the reaction stage to the greatest extent possible. However, equation 1.4 calculated the rotating row efficiency, neglecting any losses in the nozzle or stationary blade row. As with the reaction stage, the variation of efficiency varies with the ρ ratio and is influenced by the discharge angle $\alpha 1$.

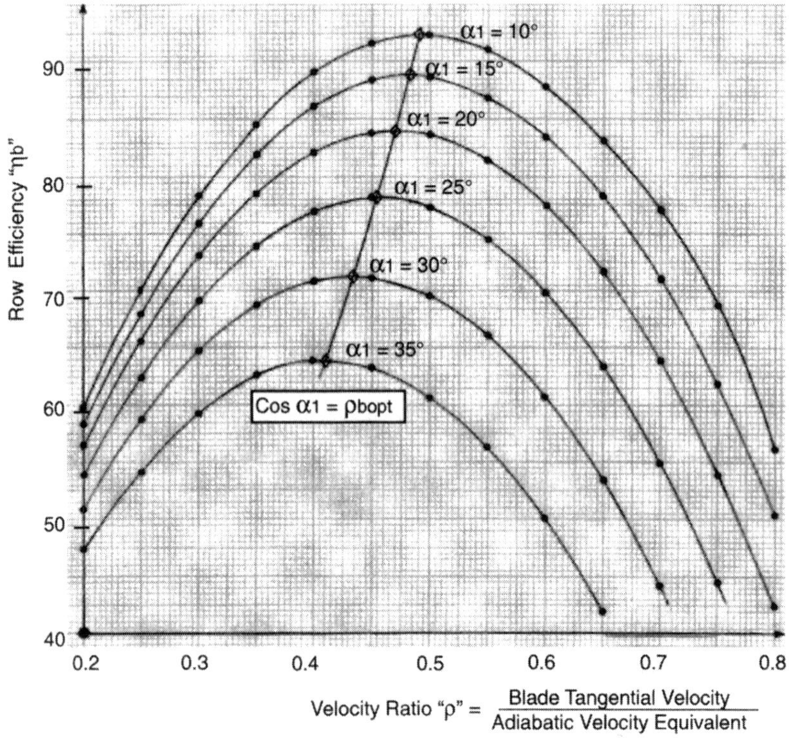

Fig. 1–14 The Efficiency of a Pure Impulse Row as a Function of the ρ Ratio for Different Steam Discharge Angles

• **Comparison of the reaction and impulse stages.** To allow a meaningful comparison of these two philosophies, stage design uses two options—one impulse and one reaction—both with a realistic discharge angle. In making the following comparison it must be considered that for the reaction stage the efficiency is that of the gross stage ηgs and for the impulse that of the blade row ηb. In Figure 1–15, this is shown with the efficiency curves for a discharge angle of 15° on each design. This angle is a realistic value for many stages. Other stage parameters have been kept as constant as realistic to allow some level of comparison.

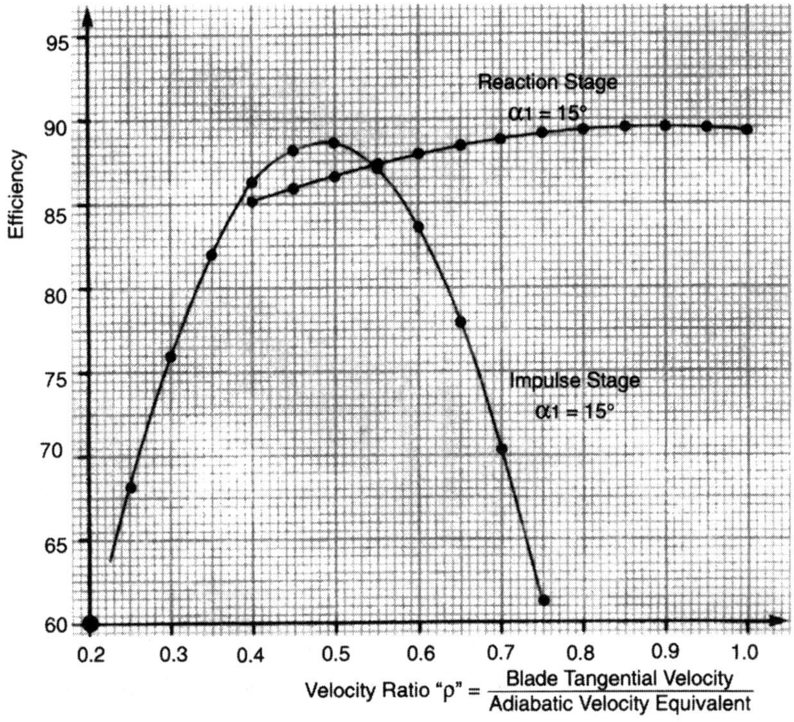

Fig. 1–15 Comparison of the Impulse and Reaction State Efficiencies as the Ratio ρ Varies

- **Impulse stage more sensitive.** Possibly the most significant fact to emerge from this comparison figure is that the impulse stage is far more sensitive to minor changes in the ρ ratio than the reaction stage. Various conditions in a unit are most likely to cause the actual ρ value to be different from the design optimum. The two major conditions are

    1. That the impulse stage as built does not produce a discharge angle in accordance with design specification. The discharge pressure from any stage is a function of the discharge area. Therefore, any deviation will change the pressure ratio on the stage, which will change the enthalpy drop, thus modifying the Isentropic Velocity Equivalent and compromising the ρ figure.

    2. That the stages have sustained some level of mechanical damage that has either opened or closed the vane discharge edges, again modifying the discharge area.

- **The degree of reaction.** Figure 1–16 is a portion of an expansion line *R-S,* with the static steam condition (without consideration of the velocity energy), at inlet to a stage shown by point *A* and at discharge by point *E.* The isentropic drop from *A* to the discharge pressure is shown as point *F.* There is also carry-in energy *ha* from the previous stage. Within this stage, the steam is to expand from pressure *pi* to *pd.* At some intermediary point, pressure *pm,* which is between these two pressures, the steam will discharge from the stationary blade row and flow into the rotating elements.

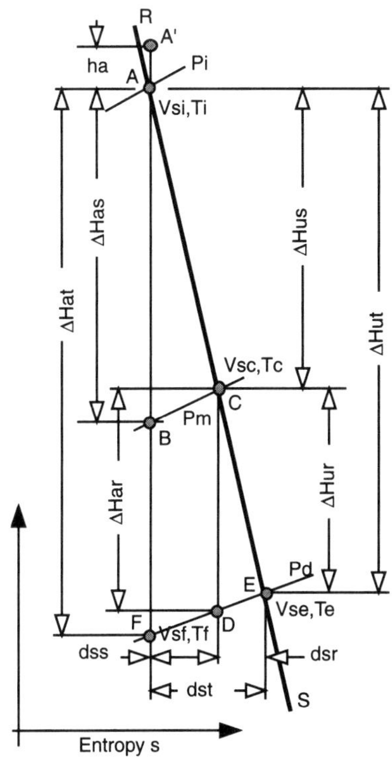

Fig. 1–16 A Stage on the Mollier Diagram

The pressure at which the steam leaves the stationary blades and flows into the rotating elements establishes the distribution of thermal energy and the pressure expansion in the stage. The ratio of isentropic energy across the rotating blades to the isentropic energy in the total stage is called the degree or percent of reaction *Rx* and can be defined mathematically as:

$$Rx = \frac{\Delta Hur}{\Delta hat} = \frac{\Delta Hur}{\Delta hus + \Delta H\,ur} \tag{1.8}$$

The degree of reaction that is selected for any stage is chosen to represent the design philosophy and experience of the manufacturer that will build the unit. (A higher degree of reaction ensures there is a positive pressure on the rotating blade at all diameters, with no possibility of negative root reaction, which produces an upstream pressure, with the pressure from the rotating blade being higher than the inlet). Having established the design degree of reaction, this will also establish the exhaust enthalpy from the stationary blades.

Therefore, the degree of reaction will also establish the pressure at the stationary/rotating blade row interface. The selection of this interface point or stationary blade discharge pressure defines the point *B* at pressure *Pm*. The intersection of this pressure locus *Pm* and the state line, point *C*, defines the steam conditions at exit from the stationary blades. At this point the steam has conditions *Pm*, *Vsc*, and *Tc*.

STATE-LINE EFFICIENCY

In the turbine, stage losses occur that are associated with expansion of the steam through the blade rows. There are a number of factors that contribute to these, but their total effect is to reduce the available or isentropic energy by a loss amount. However, the steam still issues from the blade row at the same pressure as would have occurred had the pressure drop been isentropic. The effects of these losses on the stage state line are shown in Figure 1–16.

Steam at condition *A* is admitted to the stationary blade row at a pressure *Pi*, and expands to condition *C* pressure *Pm*. At this point, the available energy ΔHas has suffered a loss so the useful energy converted to kinetic velocity is ΔHus. There is, therefore, a loss of enthalpy of ΔHas - ΔHus, and there has been an entropy increase of *dss.* At condition *C,* the steam enters the rotating blade row and again expands to condition *E* and leaves the rotating blade row at a pressure *Pd.* The available energy on the rotating row is *ΔHar,* and, due to losses, this is reduced to *ΔHur.* This loss causes an entropy increase of *dsr.*

The location of these points at one radial location (stream line), are shown in Figure 1–17, which identifies axial positions *A, C,* and *E* with pressures *Pi, Pm,* and *Pd.*

From Figure 1–16, it can be seen that the total available energy *ΔHat* is reduced to the total useful of *ΔHut,* and that there is a total entropy increase of *dst.* The state-line efficiency is defined as the ratio of the useful to the available energy. If the state-line efficiency is *ηsl,* then for this stage:

$$\eta sl = \frac{\text{Useful Energy}}{\text{Available Energy}} = \frac{AE}{AF} = \frac{\Delta Hut}{\Delta Hat}$$

This equation was seen earlier as equation 1.1. Considering this definition of efficiency, written as *ηsl,* can give in terms of the kinetic energy equivalent of both enthalpies as follows:

$$\eta_{sl} = \frac{C1^2}{C0^2}$$

Similarly, if the velocity coefficient *Φv* for the stage is defined as *C1/C0,* then the stage efficiency *ηsl* and the velocity coefficient are related by the expressions:

$$\eta sl = \Phi v, \text{ or } \Phi v = \sqrt{\eta sl} \qquad (1.9)$$

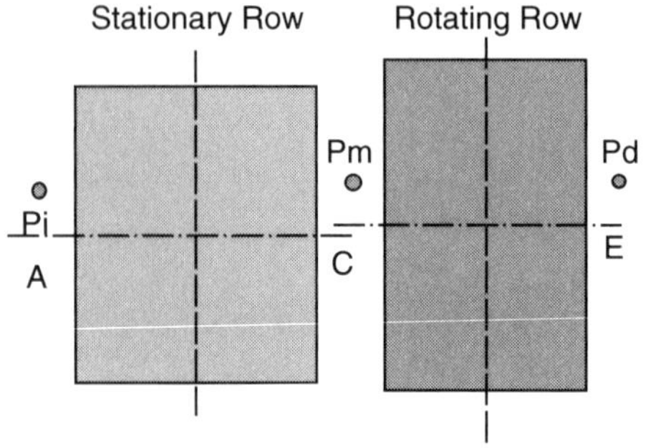

Fig. 1–17 A Stage Showing the Condition at One Radial Location

The preceding analysis of Figure 1–16 has assumed the efficiencies in the stationary and rotating blade rows are the same, and the locus *A-E* represents the change in steam conditions throughout the stage. In fact, the true efficiency of the two rows may be somewhat different because it is often possible to achieve a higher efficiency in one row of elements than the other. Under these circumstances, the true condition might be more correctly represented by the locus *A-X-C,* as shown in Figure 1–18. In this case, *A-X* represents the expansion in the stationary blades, and *X-E* the expansion in the rotating blade row. This, for most purposes, is a small effect and can be neglected. However, in an operating unit where one row of elements has sustained damage, this effect could be quite significant and has the potential to modify the steam conditions through the remainder of the section or even the entire unit. However, this cascading effect becomes less evident with continued expansion down the steam path.

In Figure 1–18, the energy available to the rotating blade row is shown as ΔHar, coming from conditions *B-F*. In fact, the energy to be expended over this row is from *X-D,* assuming the inlet conditions are represented by those at condition *X*. However, the energy range *X-D* is greater than *B-C* due to a phenomenon known as the reheat effect. The reheat effect takes account of the frictional and other energy losses that occur within a blade row, raising the temperature of the blade row metal, and then returning this energy as a heating effect on the steam flowing over these surfaces. This is evident from an examination of the Mollier Diagram, in that as the pressure lines move to the right they are convergent.

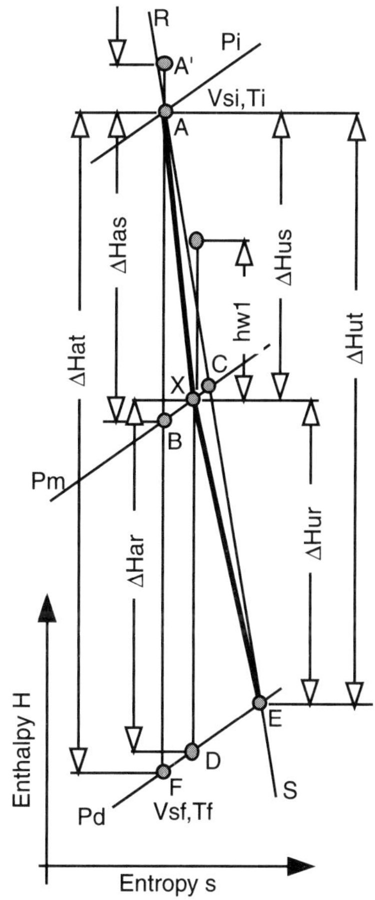

Fig. 1–18 The State Line Expansion of a Stage with Different Efficiencies in the Stationary and Rotating Rows

Using the steam properties at point *E*, a knowledge of the energy expended in the stationary blade row, and the geometry of the vanes to be used can be determined from the construction of a velocity diagram for the steam discharging from this row. The data from this diagram allows the optimum rotating blade inlet angle $\beta 1$ and velocity $W1$ to be determined in the next section. Using the same analysis for the rotating blade row, including the rotating blade energy ΔHur, the full velocity diagram for the stage can be constructed.

In determining the total energy available to the stationary and rotating blade rows, it is necessary to account for the carry-in energy shown as *ha* for the stationary row and *hw1* for the rotating row. These carry-in energies are shown in Figure 1–18.

THE VELOCITY (VECTOR) DIAGRAMS

The design information developed in the thermal analysis of a unit is sufficient that blade discharge and inlet angles and discharge areas are known. Therefore, the vanes can be selected or developed to meet these requirements.

There are two aspects of velocity diagrams that need to be considered. The first is for those stages where there is a minimal change of stream-line diameters as the steam flows between the two sets of vanes, and the second is those velocity diagrams where the stream line is expanding radially. That is, it is necessary to consider both two- and three-dimensional diagrams.

Two-dimensional considerations (pure impulse)

The turbine stage stationary blade row is selected and arranged to expand the steam and then discharge it into the following row of rotating blades. These blades are securely attached to the turbine rotor and cause it to rotate. The geometric requirements of the stationary and rotating blade profiles are conveniently established in terms of the velocity and direction of the steam entering and discharging from them.

Velocity diagrams are a convenient method for representing these velocities in a turbine stage. Consider the stationary blades shown in Figure 1–19. In this diagram, the stationary blades have an effective (or profile) discharge angle of $\alpha 1$, so the steam issues from the discharge opening (or throat) between the vanes with a velocity $C1$ at an angle $\alpha 1$. This velocity value corresponds to the velocity equivalent of the total useful energy expended in the row ΔHus, meaning account is taken of the stage losses. To this value of ΔHus must be added the contributing effect of ha, making the velocity from the stationary blade row equal to:

$$C1 = 223.7\sqrt{\Delta Hus + ha} \tag{1.10}$$

The steam discharges from the stationary row at a velocity $C1$ and enters the rotating blades, which are moving in a tangential direction at a velocity U, shown in Figure 1–19. This level of information on velocities enables that portion of the velocity triangle representing the conditions at exhaust from the stationary blades to be completed. From the velocity triangle, it is determined that the rotating blades sense a steam velocity $W1$, which is the relative inlet velocity to the blade row. This steam enters the rotating blades at an angle of $\beta 1$, which defines the inlet angle required of the rotating blade vane at that radial position. This profile angle is necessary to allow the steam jet to enter the rotating blade row with minimum shock, also called incidence.

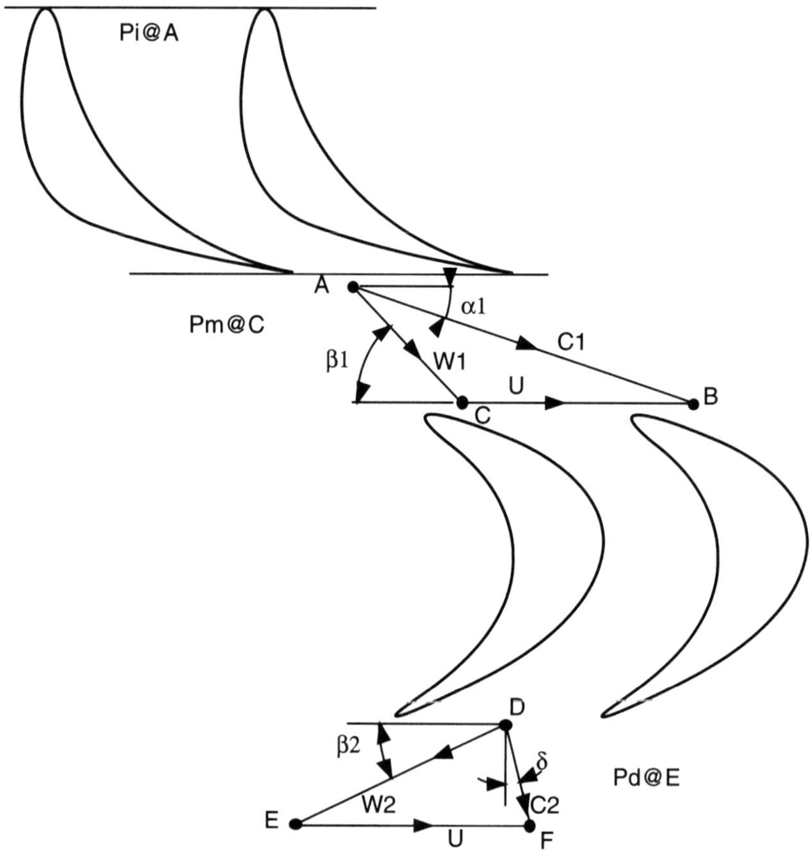

Fig. 1–19 Stage Velocity Diagram

Upon entering the rotating blade elements, the steam flows in the passage formed between the two profiles shown in Figure 1–19 and discharge from them with a velocity *W2*. This value of *W2* will be influenced by two factors:

1. The extent of aerodynamic and friction losses in the passage. The magnitude of these losses is influenced by the blade surface finish, including the effect of deposits on blade

surfaces. The discharge velocity will also be affected by any mechanical damage the profile may have sustained, causing a deterioration of the surface condition. These surface factors combine to disturb the aerodynamic flow of the working fluid and act to reduce the discharge velocity to a value that is less than the inlet velocity *W1*. That is:

$$W2 = \Phi v \,.\, W1 \qquad \textbf{(1.11)}$$

2. Any further pressure (enthalpy) drop occurred during the flow through the rotating blade passage. Such pressure drop is the reaction, which helps to increase the value of blade discharge velocity above the entry value *W1*.

$$W2 = 223.7 \times \Phi v \sqrt{\Delta Hus + hw1} \qquad (1.12)$$

These two effects will modify the velocity at discharge from the rotating blades from the inlet velocity of *W1*. Those associated with friction tend to cause a velocity reduction, while those associated with reaction increase it. These two effects must also be considered when the blade profile is selected or designed, and they must be considered in determining the size of the row discharge areas.

Figure 1–19 shows a pair of rotating blades that are receiving steam discharging from the stationary blades. This steam enters the rotating blades with a relative velocity *W1* and at an effective angle *β1*. The steam flows across the rotating blade row and discharges from it with a velocity *W2*. This velocity is modified by friction and reaction, as discussed previously, and discharges at an angle of *β2*, the designed profile discharge angle. (The actual discharge angle is a function of the ratio of blade pitch *P* to throat opening *O* rather than the physical profile angle.)

The rotating blade (because the stream line is making no significant diameter increase) still has an equivalent linear velocity of *U*. From the velocity triangle of Figure 1–19, it can be seen the steam has an absolute discharge velocity of *C2* and an absolute discharge angle of $\alpha 2$. These two parameters, *W2* and $\alpha 2$, determine the requirements of the following stationary blade row, which must accept this discharging steam. It can also be seen that the steam leaves the rotating blade row at an angle δ to the axial direction where $\delta = 90 - \alpha 2$. In undertaking the design, efforts are made to select rotating profiles so that δ is as small as possible, which means that the maximum kinetic energy has been removed from the steam.

The velocities of the rotating blades *U* at inlet to and discharge from the stage are, if no large wall-coning angle exists, of the same magnitude. In such a situation, the two velocity triangles at inlet to and discharge from the stage can be combined, as shown in Figure 1–20. In this combined diagram, additional values have been indicated. These include *Cax1* and *Cax2*, which are the axial components of the steam velocity at entry to and discharge from the stage. The change in axial velocity ΔCax is also shown.

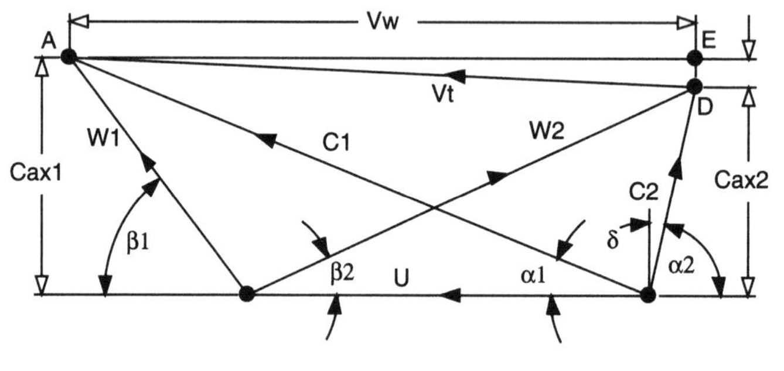

Fig. 1–20 Velocity Diagram Combining the Stationary and Rotating Diagrams from Figure 1–19

Also shown is the parameter *Vw*, which is the change of steam velocity in the tangential direction.

The thrust that is developed on the blade due to change of velocity is proportion to this total velocity change, which is proportional to *Vt*. This velocity can be resolved into two components, as shown in Figure 1–21, one in the tangential direction *Vw* and one in the axial direction. These velocity changes produce thrust on the blade, which again is resolved in two directions with magnitude proportional to the steam flow quantity in the stage *m*.

Figure 1–21 is the force diagram on the blade. This total force *Ft*, is equal to *mAD* which in this figure has been resolved into two components—one proportional to the axial change of velocity and the other equal to the tangential change.

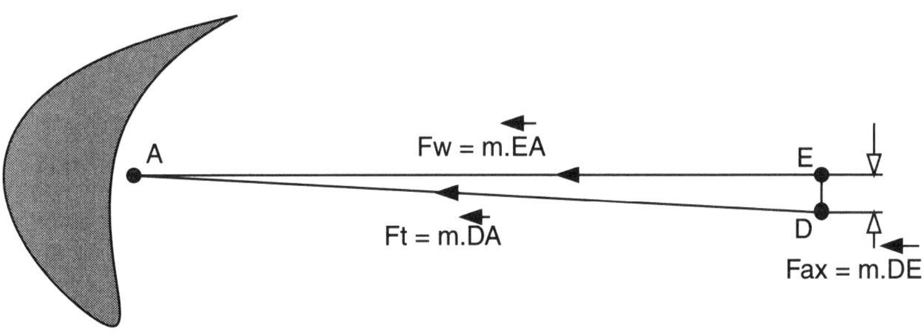

Fig. 1–21 Force Diagram for a Rotating Blade

In the tangential direction, the thrust is equal to *mAE,* which drives the blade to produce a force *Fw* in the stage. In the axial direction, the change in axial velocity ΔCax = ED produces an axial force or thrust of *Fax*, which is equal to *mED* in magnitude and direction.

Therefore:

$$Fw = \dot{m}\ \overline{AE} \tag{1.13}$$

$$Fax = \dot{m}\,\overline{ED} \quad (1.14)$$

The kinetic energy of the steam as it enters the blade row is $C1^2/2g$ per unit mass of steam. Similarly, at exit from the row, its kinetic energy is equal to $C2^2/2g$ per unit mass. Therefore, the work done on the blade is:

$$= \frac{1}{2Ug} \cdot \left[C_1^2 - C_2^2 \right] \quad (1.15)$$

At entry to the blade row, the steam has a tangential velocity of $C1.Cos\ \alpha 1$, in the direction of rotation of the blades. At discharge from the row, this tangential velocity in the same direction is $-C2.Cos\ \alpha 2$. Therefore, the change of momentum per unit mass equals:

$$C1 . \cos \alpha 1 + C2 . \cos \alpha 2 \quad (1.16)$$

Therefore, the energy given up to the blade per unit mass is:

$$\frac{U}{g} \cdot \left[C1 . \cos \alpha 1 + C2 . \cos \alpha 2 \right] \quad (1.17)$$

Therefore, equating 1.15 and 1.17 gives:

$$\frac{1}{2g} \cdot \left[C1^2 - C2^2 \right] = \left[C1 . \cos \alpha 1 + C2 . \cos \alpha 2 \right] \quad (1.18)$$

The energy supplied per unit mass of steam is equal to the kinetic energy of the steam entering the rotating blade row, and is equal to:

$$C1^2/2g \quad (1.19)$$

The blade efficiency ηb can be defined as the ratio of work done per second to the energy supplied per second. That is:

$$\eta_b = \frac{\frac{U}{g}.\left[C1.\cos\alpha 1 + C2.\cos\alpha 2\right]}{\frac{C1^2}{2g}} \tag{1.20}$$

However, from Figure 1–20, (C1.Cos α1 + C2.Cos α2) = Vw, where *Vw* is the velocity of whirl. Therefore:

$$\eta_b = \frac{2.U.Vw}{C1^2} \tag{1.21}$$

Similarly, the change of momentum in the axial direction is:

$$C1 . \sin\alpha 1 - C2 . \sin\alpha 2 \tag{1.22}$$

But, since C1.Sin β1 = W1.Sin β2, the total thrust per unit mass is given by:

$$\frac{1}{g}.\left[W1.\sin\beta 1 - W2.\sin\beta 2\right] \tag{1.23}$$

This axial thrust produces an axial force that is non productive within the steam path. It also produces an axial force that must be balanced within the steam turbine or in residual load carried by the axial thrust bearing.

Two-dimensional considerations (with high reaction)

Now consider the vector diagram for a stage with high levels of reaction. A normal definition of high reaction is when the total enthalpy/pressure drop is equally divided between the two rows. Such a stage is termed a 50% reaction design. In this design, the pressure drop in the stationary blade row is significantly reduced.

The degree of reaction *Rx* is defined as the ratio of the rotating blade enthalpy drop to the total stage enthalpy drop. An alternate definition of percentage reaction is provided in terms of the stage pressure drops, and can be quantified in the following manner. If *Pi* is the stage inlet pressure, as shown in Figures 1–16 and 1–18, *Pm* the pressure between the stationary and rotating blade rows, and *Pd* the pressure at discharge from the stage, then the degree of reaction *Rx* can be defined in terms of the stage pressures or enthalpies. Because the pressure drop in any stage is almost linear over any single stage, these two definitions, based on either pressure or enthalpy, provide substantially the same level of reaction.

Equation 1.8 provided a definition of the reaction *Rx* in terms of enthalpy drops, and the pressure gives the following equation:

$$Rx = \frac{Pm - Pd}{Pi - Pd} \tag{1.24}$$

The most meaningful way to consider the velocity diagram for a reaction stage and how the enthalpy distribution within a section affects stage geometry—specifically the blade vanes—is to make a direct comparison with that of the pure impulse stage when other stage parameters are as comparable as the design philosophy of the two allows. Consider the following examples.

Example 1 (the pure impulse stage)

This pure impulse stage on a unit operating at 3600 rpm has a mean diameter of 40.0 in. a steam discharge angle $\alpha 1$ from the stationary blade row of 12° and a velocity coefficient Φv of 0.98. The rotating profile discharge angle is 23°. With this information, we can make the following calculations:

U	= 15.71 x 40	= 628.4 ft/sec.
ρ optimum	= (Cos 12°)/2	= 0.489
Isentropic velocity	= 628.4/0.489	= 1284.9 ft/sec.
Isentropic enthalpy drop	= $(1284.9/223.7)^2$	= 33.00 Btu/lb
Stage efficiency	= 0.98^2	= 0.96
Useful enthalpy drop	= 0.96 x 33.00	= 31.68 Btu/lb
Nozzle velocity, (C1)	= $223.7\,(31.68)^{1/2}$	= 1259.1 ft/sec.
C1 (tangential)	= C1 . Cos 12°	= 1231.6 ft/sec.
C1 (axial)	= C1 . Sin 12°	= 261.8 ft/sec.

Using this data:

$$W1 = \sqrt{(C1tg - U)^2 + C1ax^2} \qquad = 657.6 \text{ ft/sec.}$$

$$\beta 1 = \mathrm{Tan}^{-1} \frac{C1\,ax}{C1\,tg - U} \qquad = 23.5°$$

W2	= Φv . W1	= 644.5 ft/sec.
W2 (tangential)	= W2 . Cos 23°	= 593.3 ft/sec.
W2 (axial)	= W2 . Sin 23°	= 251.8 ft/sec.

$$C2 = \sqrt{(U - W2tg)^2 + W2\,ax^2} \qquad = 254.7 \text{ ft/sec.}$$

$$\delta = \mathrm{Tan}^{-1} \frac{U - W2\,tg}{W2\,ax} \qquad = 7.9°$$

The velocity diagram for this stage is shown as Figure 1–22. From this figure, it can be seen that the velocity component Δcax contributing to axial thrust on the rotating blade is

$$\Delta Cax = C1ax - C2ax \qquad (1.25)$$

$$= 261.8 - 251.8 = 10.0 \text{ ft/sec.}$$

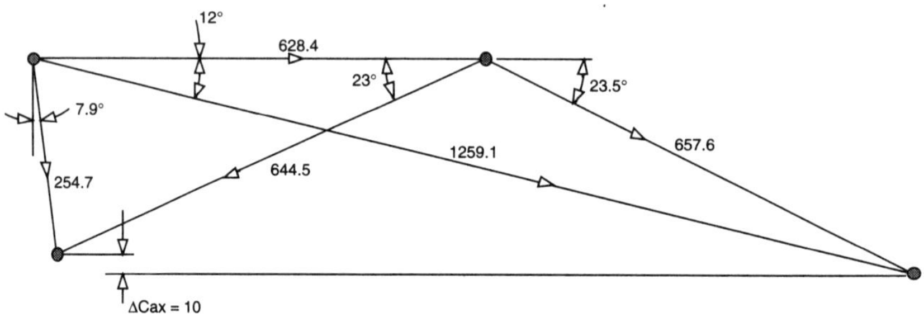

Fig. 1–22 The Velocity Diagram for the Stage as Seen in Example 1

Example 2 (a 50% reaction stage)

This 50% reaction stage on a unit operating at 3600 rpm, has a mean diameter of 40.0 in. a steam discharge angle $\alpha 1$ from the stationary blade row of 12° and a velocity coefficient Φv of 0.98. The rotating profile discharge angle is 23°. With this information, we can make the following calculations:

U	= 15.71 x 40	= 628.4 ft/sec.
ρ optimum	= (Cos 12°)	= 0.978
Isentropic velocity	= 628.4/0.978	= 642.5 ft/sec.
Isentropic enthalpy drop	= $(1284.9/223.7)^2$	= 8.25 Btu/lb
Stage efficiency	= 0.98^2	= 0.96

Useful enthalpy drop	= 0.96 x 8.25	= 7.92 Btu/lb
Nozzle velocity, (C1)	$= 223.7 \times (31.68)^{1/2}$	= 629.6 ft/sec.
C1 (tangential)	= C1 . Cos 12°	= 615.8 ft/sec.
C1 (axial)	= C1 . Sin 12°	= 130.9 ft/sec.

Using this data:

$$W1 = \sqrt{(C1tg - U)^2 + C1ax^2}$$

= 131.5 ft/sec.

$$\beta 1 = Tan^{-1} \frac{C1\,ax}{C1\,tg - U}$$

= 89.5°

W2	= Φv . (ΔW1+ΔHur)	
	= 0.98(0.34 + 7.92)= 8.09 Btu/lb	
W2 (velocity)		= 636.5 ft/sec.
W2 (tangential)	= W2 . Cos 23°	= 585.9 ft/sec.
W2 (axial)	= W2 . Sin 23°	= 248.7 ft/sec.

$$C2 = \sqrt{(U - W2tg)^2 + W2\,ax^2}$$

= 252.3 ft/sec.

$$\delta = Tan^{-1} \frac{U - W2\,tg}{W2\,ax}$$

= 9.7°

The velocity diagram for this reaction stage is shown as Figure 1–23, and the value of *ΔCax* is:

$$\Delta Cax = 130.9 - 248.7 = -117.8 \text{ ft/sec.}$$

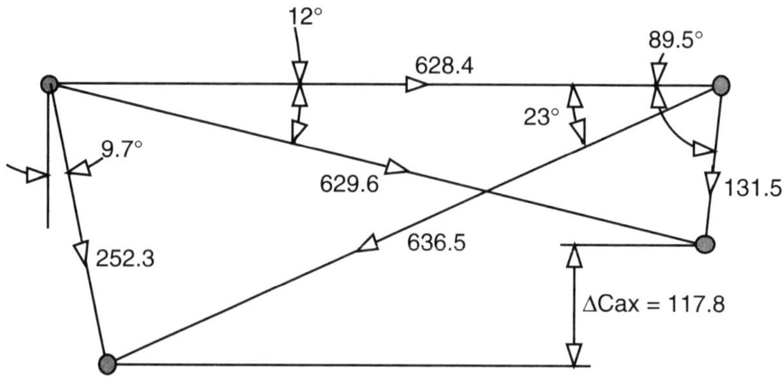

Fig. 1–23 The Velocity Diagram for the Reaction Stage

A comparison of the major parameters of these two stages is shown in Table 1–3. From this comparison, several interesting observation can be made regarding the resultant design of the impulse and reactions stages. Among these are

- The steam velocities in the reaction stage are lower than those in the impulse. One advantage of this is that the row Reynolds Numbers will be lower, and therefore it can be anticipated that the surface frictional losses will be less for any level of surface roughening.
- Both designs produce relative discharge velocities from the rotating row, which is close to axial, meaning that as much of the kinetic energy as possible will have been extracted from the steam.
- The difference in differential axial velocities produces a significantly higher reaction axial thrust in the rotating blade row. This requires a larger thrust bearing to balance the stage, if the section design is not double flowed.

Design	Impulse	Reaction
Blade tangential velocity U	628.4	628.4
Stationary discharge angle $\alpha 1$	12°	12°
ρ optimum	0.489	0.978
Steam discharge velocity C1	1259.1	629.6
Relative inlet velocity W1	657.6	131.5
Vane inlet angle $\beta 1$	23.5°	89.5°
Absolute discharge velocity W2	644.5	636.5
Vane discharge angle $\beta 2$	23°	23°
Absolute discharge velocity C2	254.7	252.3
Discharge angel to axial δ	7.9°	9.7°
Axial component of residual velocity ΔCax	10.0	-117.8

Table 1–3 Summary Comparison of the Stages

Three-dimensional considerations. The two velocity diagrams considered previously assume that diameters of the stationary and rotating blade row are sufficiently close that in preparing the velocity diagrams the stationary and rotating rows can be considered to have the same value of blade tangential velocity *U*. For short radial-height blades with no significant coning of the outer side wall, this is a justified assumption and introduces no noticeable error into the design of the stage. However, when there is a significant increase in the specific volume in the flow, it becomes necessary to increase the flow area by coning the steam path side walls to accommodate the increase in volumetric flow.

Under these circumstances of high radial flow, it is no longer possible to use the common value of *U* without introducing error into the stage sizing calculations. There is a radial flow component to the steam flow that must be considered in making the stage design.

Figure 1–24 shows a three-dimensional velocity diagram for a stage with a large radial flow component. In this diagram the following nomenclature is used:

Stationary Blade Row:

- Ust relative blade velocity at nozzle discharge radial position
- C1 steam discharge velocity in the axial plane
- W1 steam relative velocity in the axial plane
- $\alpha 1$ steam discharge angle from stationary blade row
- $\beta 1$ required rotating vane inlet angle
- Cax1 axial component of steam velocity from the stationary blade row
- C1′ absolute steam discharge velocity from stationary blade row
- W1′ steam relative velocity from the stationary blade row
- $\gamma 1$ steam line inclination at discharge from the stationary blade row
- Cr1 steam radial flow component

Rotating Blade Row:

- Urot rotating blade tangential velocity at inlet radial position
- W2 relative steam discharge velocity in the axial direction
- C2 absolute steam discharge velocity in the axial direction
- $\beta 2$ rotating blade discharge angle
- $\alpha 2$ absolute steam discharge angle
- Cax2 axial component of steam velocity from the rotating blade row
- W2′ relative steam velocity at discharge from the row
- C2′ absolute steam velocity at discharge from the row
- $\gamma 2$ stream line inclination at discharge from the rotating blade row

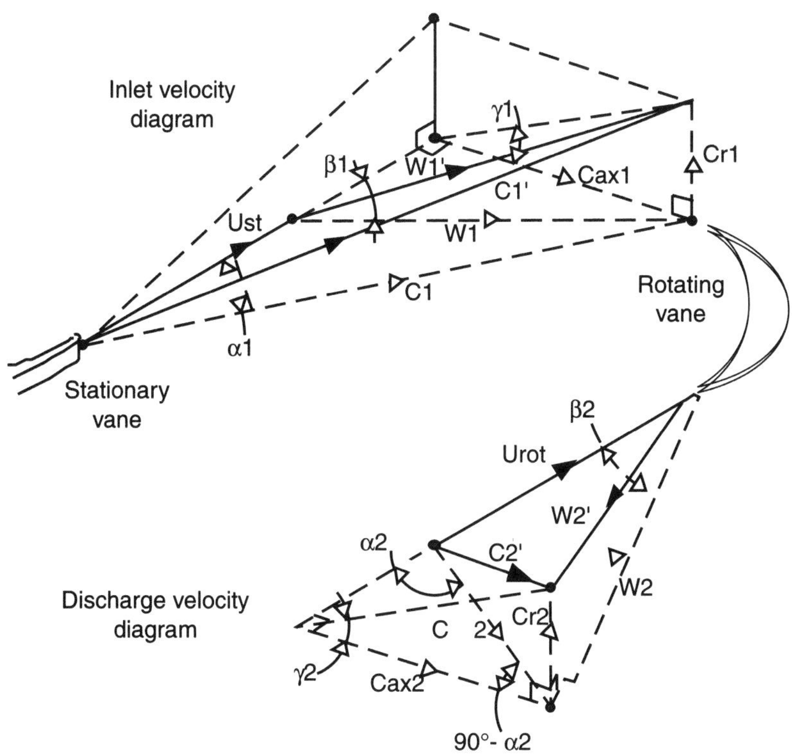

Fig. 1–24 Three-Dimensional Velocity Diagram

Figure 1–24 is a three-dimensional velocity diagram as required for stages with a large radial flow component. To calculate this, figure the steam velocity from the stationary blade row is *C1′* equivalent to *Δhus*, which is then resolved into the components shown. The rotating row is calculated similarly. The relative velocity leaving is *W2′*, which again can be resolved into the component velocities shown.

These velocity diagrams are solved by triangulation as in the two-dimensional types. In this diagram, the velocity discharging from the stationary bade row *C1′* is determined from the enthalpy drop *ΔHus*. Similarly, the relative velocity from the rotating row *W2′* is found as the sum of the enthalpy equivalent to *W1′* and *Δhur*.

STEAM PATH SIZING AND ARRANGEMENT

The steam path components must be sized and arranged so that they achieve certain requirements. These are

- To pass the correct quantity of steam. The heat balance defines the quantity of steam required to produce the output required from the turbine. The blades must be of sufficient size they are able to pass the flow and not exceed axial velocities limits as defined by design
- Be so arranged that the optimum ratio ρ is achieved
- The components of the stage must be arranged on diameters that the lap L as discussed later and shown in Figures 1–30 and 1–31 are achieved at the inner and outer blade positions

Manufacturers tend to have available certain designs of control stages which are used for their nozzle-controlled units. These stages are normally subjected to high levels of dynamic loading, most particularly at part load when the stage is receiving steam from partial arc admission.

For the purpose of illustrating methodology, assume an impulse stage with a steam flow to the control stage, after valve leakage, of 1,842,036 lb/hr being admitted to the unit over a 95% arc of the control stage stationary blade row. (The 5% is the blank space between the active nozzle arcs.) The stage is then designed for a flow of 1,842,036/0.95 = 1,938,985lb/hr over a 100% admission arc. Also assume that the steam has a specific volume of 0.3933 cu. ft./lb, and the nozzle has a discharge angle of 12°, and the stationary nozzles are mounted at a mean diameter of 38.50 in.

Solution of vane heights. If the velocity triangle is solved for a useful enthalpy drop of 26.08 BTU/lb, then Figure 1–25 is produced and the following solution to vane heights can be established:

The nozzle discharge height *Hn* is:

$$Hn = \frac{1{,}938{,}985 \times 0.3933}{25\pi \times 38.5 \times 1{,}142.2 \times \text{Sin } 12^\circ}$$

$$Hn = 1.062 \text{ in. say } 1.06 \text{ in.}$$

The nozzle discharge area *An*:

$$= \pi \, . \, Dm \, . \, Hn \, . \, \text{Sin } \alpha 1$$

$$= \pi \times 38.5 \times 1.06 \times \text{Sin } 12^\circ$$

$$= 26.66 \text{ sq. in.}$$

However, this represents the area requirements for 360° of admission, and a flow of 1,938,985 lb/hr. For a 95% admission arc, the actual area required is 26.66 x 0.95 sq. in. This gives:

$$An = 0.95 \times 26.66 = 25.33 \text{ sq. in.}$$

The rotating blade will require a lap sufficient to allow the steam to flow from the stationary blade to the rotating, minimizing spillage over the tip or coverband. Assume a total lap (inner plus outer) of approximately 0.15 in. to 0.17 in. is acceptable.

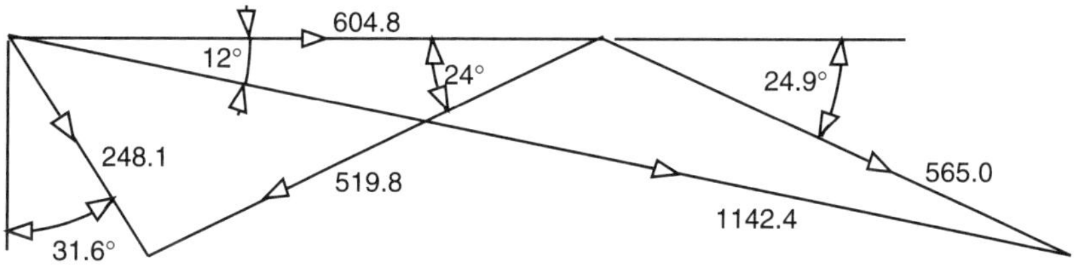

Fig. 1–25 The Control Stage Velocity Diagram

Note that the lap required for any stage is selected by the designer on the basis of his experience in the amount of radial overlap required to help ensure the steam discharging from the stationary blade row is directed into the rotating while considering the effects of radial flow in the axial gap between the stages. The total lap is normally arranged so that about 66 to 75% of the total is placed at the tip location of the stage.

Therefore, the rotating blade height *Hb* should be between:

$$1.06 + 0.15 = 1.21 \text{ in. and } 1.06 + 0.17 = 1.23 \text{ in.}$$

To provide the outer lap with a larger amount of the lap than the inner, the mean diameter of the rotating blade row is made larger than that of the stationary row. Therefore assume a rotating blade row of mean diameter:

$$38.50 + 0.04 = 38.54 \text{ in.}$$

Note that the 0.04 in. increase in mean diameter of the rotating blade row is again a design selected value which reflects the designer's opinion of the diameter difference together with the lap required to accommodate radial flow by achieving an acceptable split of the lap.

Therefore, if we assume a rotating blade row total lap of 0.15 in., then the rotating blade has a radial height of 1.21 in. and root and tip diameters of:

row root diameter	$Dr = 38.54 - 1.21$	$= 37.33$ in.
row tip diameter	$Dt = 38.54 + 1.21$	$= 39.75$ in.
the inner lap =	$\{(38.50 - 1.06) - 37.33\}/2$	$= 0.055$ in.
the outer lap =	$\{39.75 - (38.5 + 1.06)\}/2$	$= 0.095$ in.
blade discharge area	$Ab = \pi \cdot Dm \cdot Hb \cdot \sin\beta 2$	
	$= \pi \times 38.54 \times 1.21 \times \sin 24°$	
	$Ab = 59.59$ sq. in.	

This stage radial layout of this control stage is shown as Figure 1–26. The axial widths of the blades are estimated at this time, with the final width requirements determined from a mechanical analysis of the stage.

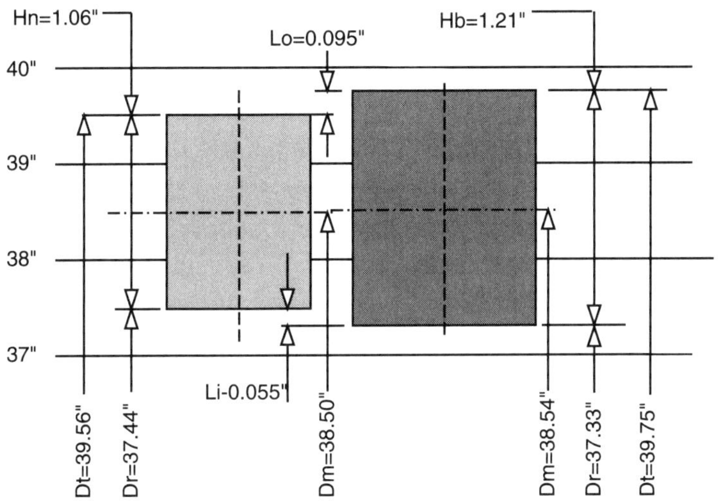

Fig. 1–26 Principle Radial Dimensions of the Control Stage Stationary and Rotating Blade Rows

Assume this design requires a small degree of reaction at the root of these stages with a velocity ratio ρ of 0.55 at the inner diameter, and the available enthalpy for the high-pressure section from the heat balance after the control stage is 151.5 Btu/lb. How can this be divided among a number of stages?

For any series of stages, having the same ρ value, the first stage in the group will have an enthalpy $\delta h1$ at the design diameter of:

$$\delta h1 = Dm1^2 \cdot \frac{\Delta H\,(\text{useful})}{\Sigma\ Dm^2} \qquad (1.26a)$$

where:

$\delta h1$ is the first stage enthalpy

Dm1 is the mean diameter of the first stage

ΔH is the total heat drop on the section, including reheat

Σdm is the sum of the individual stage mean diameters squared

A more general form of this equation to find the enthalpy drop on any stage *n* in the group or section is

$$\delta hn = \Delta Had \, . \, \frac{Dmn^2}{\Sigma \, Dm^2} \, , \text{ or } \quad \delta hn = \Delta H \, . \, \frac{Dmn^2}{\Sigma \, Dm^2} \qquad (1.26b)$$

where:

δhn is the energy on stage *n*

ΔHad or ΔH is the total energy on the section (either adiabatic or useful)

Dmn is the mean diameter of stage *n*

ΣDm^2 is the square of the mean diameters of all stages

There is a series of combinations of stages and root diameters that could be used to satisfy the requirement of a ρ value of 0.55 at the root. The design engineer investigates these and identifies the most suitable. Alternative stage arrangements for a 3600 rpm unit are shown in Table 1–4.

Number of Stages	**5**	**6**	**7**	**8**
Alternate	{a}	{b}	{c}	{d}
Stage Ho Btu/lb	151.4 ÷ 5	151.4 ÷ 6	151.4 ÷ 7	151.4 ÷ 8
Btu/lb/Stage	30.28	25.23	21.63	18.93
Co ft./sec.	1,231.0	1,123.7	1,040.4	973.2
		Velocity Ratio U/Co = 0.550		
Ur inches ft./sec.	677.0	618.0	572.2	535.2
Dr inches	43.12	39.37	36.44	34.09

Table 1–4 Possible Section Configurations

Figure 1–27 shows the layout of alternates *b, c,* and *d* after the control stage. These represent arrangements of alternates *b, c,* and *d* from Table 1-4. From this figure alternate *c* would appear to provide a suitable arrangement. Using the methodology discussed previously for sizing the vane heights, a layout of the complete steam path is shown as Figure 1–28.

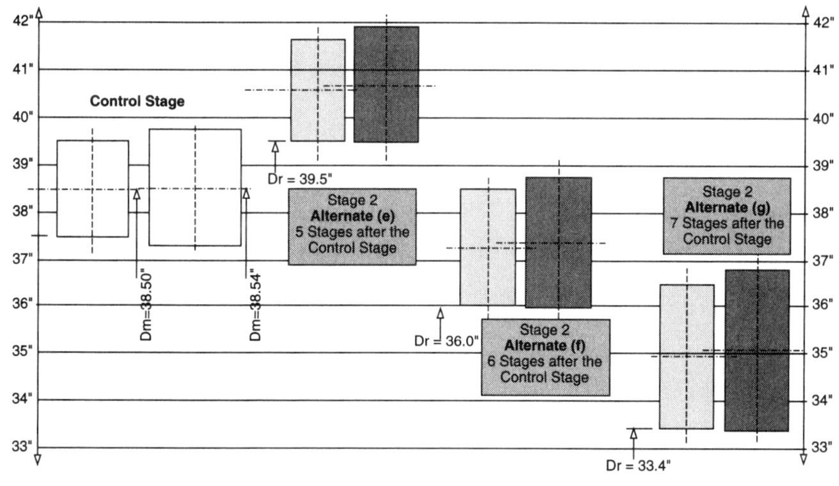

Fig. 1–27 Alternate Arrangements of Stage 2 Relative to the Control Stage

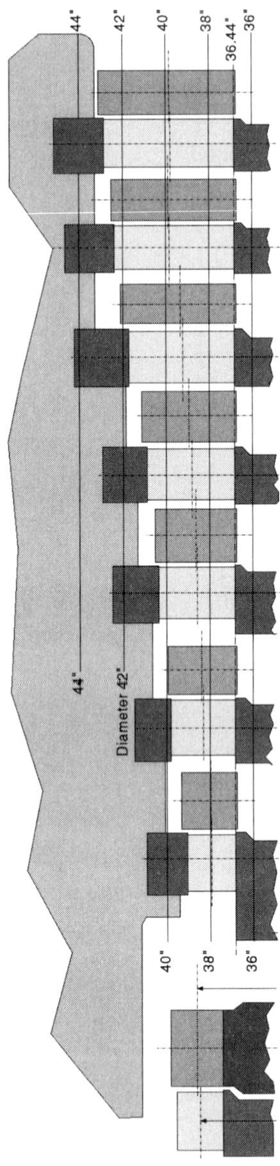

Fig. 1–28 The Calculated Steam Path, Including the Control Stage of Figure 1–26 and the 6 Impulse Stages in Table 1–4

RADIAL PRESSURE GRADIENT

The effect of this radial outward flow of steam is to increase the pressure from root to tip sections in the axial gap between the stationary and rotating blade rows. The resultant effect of this phenomenon is for the enthalpy level to rise in the axial gap with the pressure increasing toward the tip. This results in a reduction of enthalpy in the stationary blade row towards the tip section and an increase in the rotating row.

A convenient means of determining the degree of reaction at one radial location relative to another is by this simple equation.

$$\frac{R_x}{R_y} = \left\{\frac{D_y}{D_x}\right\}^2 \qquad (1.27)$$

where:

Rx is the reaction at diameter Dx

Ry is the reaction at diameter Dy

This equation serves well for the shorter radial height stages but becomes less accurate as the vane length increases. With a ratio of height/mean diameter greater than about 20 to 25%, it is best to employ a radial pressure gradient calculation.

Using equation 1.27, the reaction at any radial location can be determined in terms of the known reaction at any known diametral position in the stage. A representation of the change of enthalpy in a stage is shown in Figure 1–29, portraying both the impulse and reaction stages.

Figure 1–29a shows the enthalpy distribution in an impulse stage, where a root reaction is assumed. Similarly, the variation of stage reaction is shown in Figure 1–29b for a reaction stage where the reaction at the mean diameter is known.

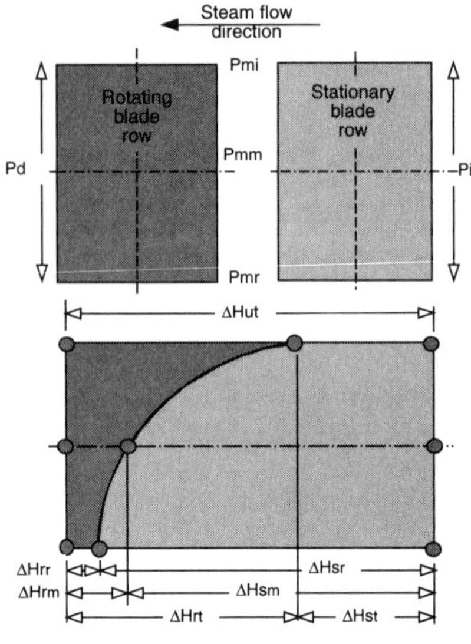

Impulse Stage

Fig. 1–29a Enthalpy Distribution between the Stationary and Rotating Blades in an Impulse Design Unit

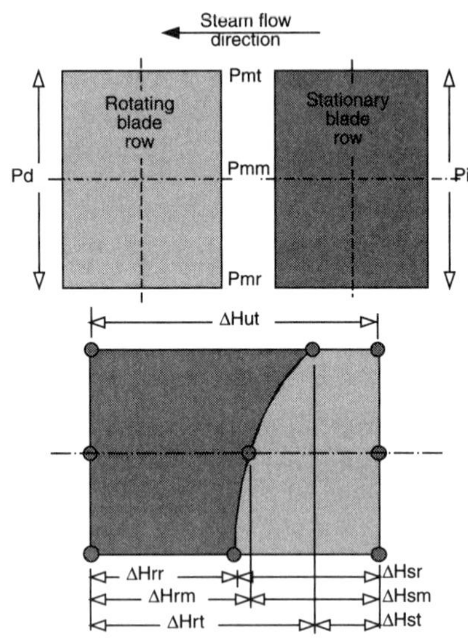

Reaction Stage

Fig. 1–29b Enthalpy Distribution between Stationary and Rotating Blades in a 50% Reaction Design Unit

Typical reaction levels in a reaction stage are 50% at the mean diameter, and in the impulse stage design is at the 5% level at the root diameter. From this information the reaction at any other radial location can be determined.

STAGE CONSTRUCTION DETAILS

The two design philosophies—impulse and reaction—result in a stage layout that is suited to their manner of energy release and the means of optimizing the efficiency of the turbine section. However, there are distinct differences in the general appearance of these two stages, as shown for the impulse design utilizing a wheel and diaphragm construction in Figure 1–30 and for the reaction stage as shown in Figure 1–31.

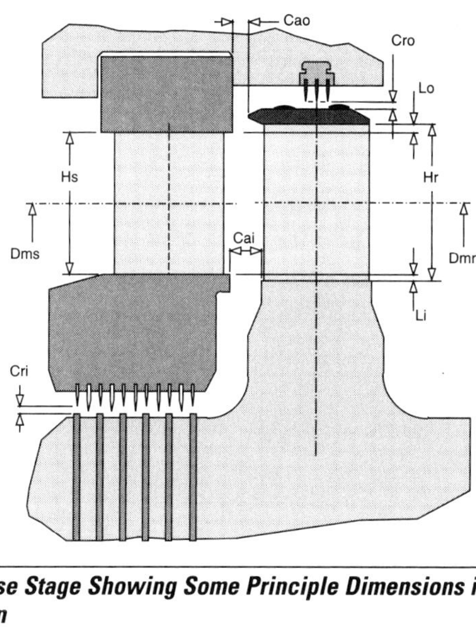

Fig. 1–30 The Impulse Stage Showing Some Principle Dimensions in the Cold Stationary Condition

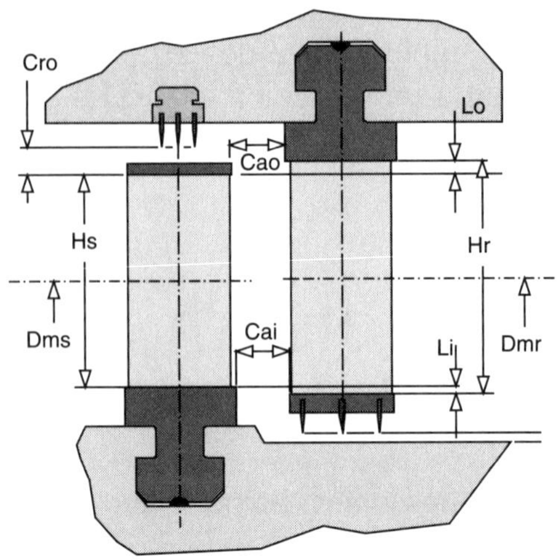

Fig. 1–31 The Reaction Stage Showing Some Principle Dimensions in the Cold Stationary Condition.

These two figures show a number of stage characteristics that are critical to the efficient and reliable operation of the stage, including:

The stage diameters (D). The diameters establish the value of the blade tangential velocity so that the enthalpy drop in the stage can be established to optimize stage efficiency.

The diameters are also selected so that under the influence of radial stretch during operation due to both temperature and stress growth, they will remain aligned with the stationary blade row.

The blade discharge heights (H). The discharge height is selected so that with the mean throat that is produced at the vane discharge edges the correct stage discharge area is achieved. This area establishes the pressure at row discharge that in turn defines the enthalpy.

The stage laps (L). The laps are the diametral differences in radial position at discharge from one stage and entry to the next. These are important parameters and are arranged so that the lap at the tip section is 60 to 75% of the total at that position.

The radial clearances (Cr). The radial clearances are selected and set in the cold stationary condition so that when hot and rotating they maintain a radial clearance between rotating and stationary components that will not generate excessive heat if rubs occur.

In addition to the more common labyrinth seals, there are other designs incorporating honeycomb systems designed so that the rotating portion of the blade row can cut a clearance into the seal. This minimizes the leakage which will occur.

The axial clearances (Ca). Like the radial seals, the axial seal is set in the cold stationary condition so it will be most effective when the unit is in operation. Unfortunately, the axial seal is often subjected to rubs that can be severed under the conditions of excessive differential expansion between the rotors and stationary portions of the unit.

FEED WATER HEATING TRAINS

The number, type, and arrangement of feed heaters in any installation is a matter for evaluation by the owner and/or the architect engineer. The final selection is normally based on the level of thermal gains that can be achieved by any modification or addition, the cost of fuel, predicted load factor, and the value of incremental output to the operator.

The number of heaters used affects cycle efficiency—the greater the number of heaters the higher the efficiency. However, the number of heaters that can be reasonably used is limited because heaters are expensive and the number of locations in the steam path where steam can reasonably be extracted is limited. These considerations, and other factors, tend to optimize the number that can be economically employed. There is also the law of diminishing returns, in which increases beyond a certain number producing only marginal gains. The architect engineer will normally investigate the number of heaters, their type, and terminal temperature differences in selecting the cycle for any installation.

There are some basic considerations that influence the selection and arrangement of heaters, but no definitive rules are available. The number of heaters normally selected being a function of the initial steam conditions, the output of the cycle, and fuel costs. These suggest some basic arrangements of heaters used and found to be convenient in modern plants. However, no indication was given of the heater type or their possible arrangement. In selecting heater configuration, the following factors need to be considered:

- The number of heaters can be increased with economic justification as output, steam conditions, and fuel costs increase. To the extent possible, anticipated fuel costs should be factored into the initial design of the plant, as the retrofitting of heaters is not possible in response to changing fuel costs.
- The bottom heater should utilize steam extracted at as low a pressure as possible—preferably extracted from just ahead of the L-0 stationary blade row.
- The deaerating heater should utilize steam (at full load) that is above atmospheric by a factor sufficient to ensure above atmospheric pressure in the vessel at all loads.
- The individual heater temperature increases should be as even as possible This is controlled by the extraction points available within the steam path and the temperature at these locations. The one exception to this is the extraction from the cold reheat, which is on average about 1.5 times the average high-pressure heater rise. Therefore, the total temperature rise through the feed train should be divided between the heaters as evenly as possible, and the individual rises should preferably not exceed 80°F. However, this value can sometimes be modified by plant economics and unit arrangement.

- On a cost basis, the number of high- and low-pressure heaters should be selected to make the number of high-pressure heaters lower than the number of low-pressure elements. It is normally more expensive to purchase and maintain high-pressure heaters. Often, on larger installations where the extraction flows are large and the extracted steam has a higher specific volume, it becomes necessary to parallel flow low-pressure heaters, because of the volumes involved.
- The top heater must raise the temperature of the feed water to, or close to the FFWT specified by design. The FFWT is normally defined so the heat added in the boiler is only the superheat portion at the local pressure. Therefore, the boiler is not required to add any, or only a minimal amount of, latent heat of evaporation to the feed water.

Shown as Figure 1–32 is a heater train in which there are six feed heaters. The train is arranged so there are two high-pressure heaters *A* and *B*, a deaerator *C* and three low-pressure elements *D, E,* and *F.* The flows into each of these heaters are shown as *Qa . . . Qf,* and the flow from the turbine exhaust is indicated at *Qt.* Each of the non-contact heaters has a drain cooling section and a 5°F temperature differential at inlet. Without information on the quantities of steam *Q* and their heat content it is not possible to complete a heat balance around them. However, it is possible to trace the flows *Q* in this train. For simplification, any heat or mass transfer resulting from extraneous flows and secondary heaters has been omitted. However the cascaded drains from heaters *D, E,* and *F* are sent to the condenser, and together with the turbine exhaust flow, form quantity *Qm1* which goes through the condensate extraction pump (CEP) to the bottom heater.

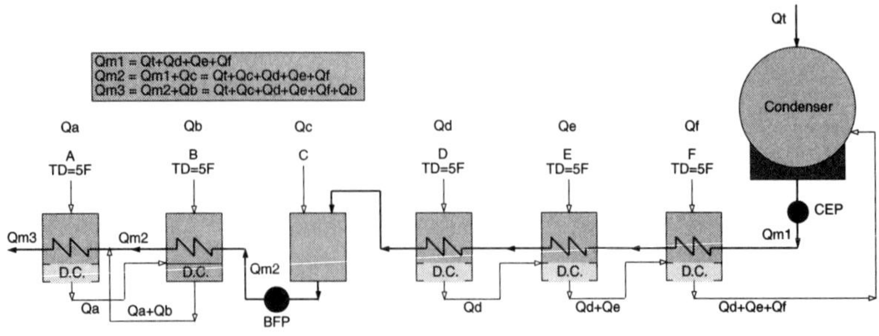

Fig. 1–32 A 6-Heater Train Comprising 3 Low-Pressure Heaters, a Deaerator, and 2 High-Pressure Heaters

The feed water flow *Qm1* passes through the three heaters *F, E,* and *D* and is then sprayed into the deaerator *C* where it mixes (makes direct contact) with turbine extraction quantity *Qc*. The resulting flow Qm2 = (Qm1 + Qc) is then pumped by the boiler feed pump to heater *B* where it is heated by the turbine extraction quantity *Qb* and the drains from heater *A* quantity *Qa*. The total condensate in heater *B* is then pumped forward into the feed line, increasing quantity *Qm2* to *Qm3* by the addition of *Qa+Qb*. The quantity *Qm3* will then be passed to the boiler where heat is added, raising its temperature to the turbine inlet condition. This analysis neglects other secondary flows, system leakages, and make-up requirements.

Shown in Figure 1–33 are the arrangement, flows, and thermal conditions around a seven heater cycle for a 140,000 kW fossil reheat unit with initial and reheat steam condition of 2415psia/ 1000/1000°F, and an exhaust pressure of 1.0 in. Hga. The FFWT is 470°F. The heater train has two high-pressure heaters, *A* and *B*, a deaerator *C* and four low-pressure elements *D, E, F,* and *G*. The flows and thermal conditions around these heaters are shown in Figure 1–33a, and the gradients of temperature, enthalpy, and flow quantity are shown in Figure 1–33b. This section also displays the thermal gradients and enthalpy rises in the train to the final feed water condition of 470.0°F. In this train, the high-pressure heaters are cascaded to the deaerator and the three highest of the low-pressure heaters are cascaded to the bottom heater, whose drains are pumped into the feed water line ahead of the second heater *F*.

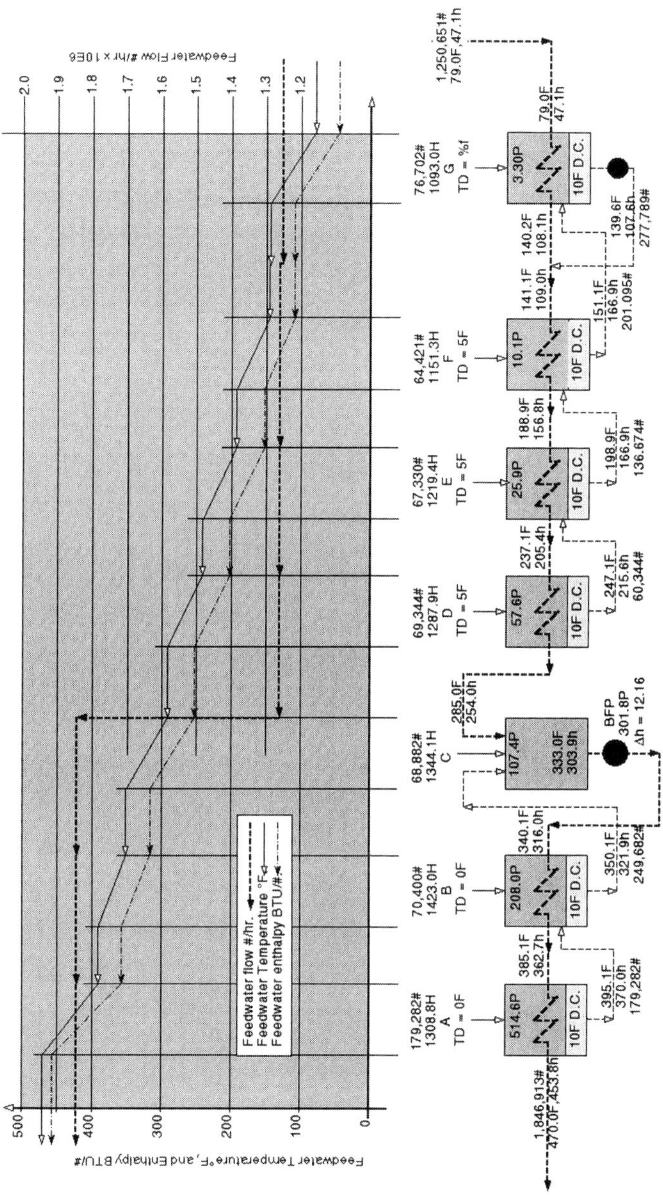

Fig. 1–33 A 7-Heater Feed Train, Showing Thermal Conditions and Flows at Each Heater

Table 1–5 shows the thermodynamic requirements in terms of flow and thermal characteristics of the steam around the heaters. In terms of the requirements this places on the steam turbine, this steam must be removed first from the steam path and then through an outer and possibly an inner casing or between blade carriers. In the case of multiple flows, there are often requirements for symmetrical extractions from both ends to maintain even axial thrusts, and when more than one double-flow section in the low-pressure sections is used, there is often a requirement of balancing the flows through the last stage blades to maintain blade loading at acceptable levels.

Heater:	G	Mixing	F	E	D	C	BFP	B	A
Temperature at Inlet:	79.0	140.2	141.1	188.9	237.1	285.0	333.0	340.1	385.1
Temperature at Discharge:	140.2	141.1	188.9	237.1	285.0	333.0	340.1*	385.1	470.0
ΔT though equipment:	61.2	0.9	47.8	48.2	47.9	48.0	7.1	45.0	84.9
Enthalpy at Inlet:	47.1	108.1	109.0	156.8	205.4	254.0	303.9	316.0	362.7
Enthalpy at Discharge:	108.1	109.0	156.8	205.4	254.0	303.9	316.0	362.7	453.8
ΔH through equipment:	61.0	0.9	47.8	48.6	48.6	49.9	12.1	46.7	91.1
Feed Water Flow 1,000lb:	1,250.6	-	1,528.4	1,528.4	1,528.4	1,846.9	-	1,846.9	1,846.9
Heater Flow 1,000lb:	76.7	-	64.4	67.3	69.3	68.9	-	70.4	179.3
Drains Flow 1,000lb:	277.8	-	201.1	136.7	69.3	1,846.9	-	249.7	179.3
Drains Temperature:	145.1	-	151.1	198.9	247.1	333.0	-	350.1	395.1

Notes: Pressure in psia, temperature in °F, and enthalpy in BTU/lb.

Table 1–5 Thermal Conditions in the Heater Train

As the steam expands, the specific volume increases, until at the low-pressure end the volumetric flows can become particularly large requiring a number of pipes to remove the steam and maintain acceptable velocities in the pipes. It is normal for the architect engineer to specify the maximum pressure drop allowable in the extraction lines, and therefore the turbine designer must calculate the number and size of the extraction pipes so as not to exceed this pressure drop.

Table 1–6 shows factors that influence the sizing of the extraction lines which are to be used in any configuration. Consider the extraction to heater *A* at a pressure of 530.5 psia. Here the steam can be removed through a single line with an internal diameter of 8.24 in. However, because of the pressure, this will need to be produced as a thick-walled, high-quality line of superior material. Similarly, the extraction for heater *G* presents other problems, because the volumetric flow has increased to 7,789,000 cu.ft/hr, requiring a pipe area of 2077 sq. in. to maintain the flow at a velocity of 150 ft/sec. In fact, it would be normal to remove this flow in a number of parallel lines to maintain steam path symmetry and make the lines of a manageable diameter. Shown in Table 1–6 are a number of combinations of line sizes from one to eight that could be used in the case of a four-flow exhaust.

For heater	G	F	E	D	C	B	A
Extraction Press	3.51	10.8	27.6	61.2	114.3	221.3	530.5
Extraction Enthalpy	1093.0	1151.3	1219.4	1287.9	1344.1	1423.0	1308.8
Extraction Flow, 1000lb	76.7	64.4	67.3	69.3	68.9	70.4	179.3
Steam Specific Vol., cu ft/lb	00.5	36.5	17.5	9.23	5.595	3.330	1.116
Moisture content	0.968	1.000	1.000	1.000	1.000	1.000	1.000
Volume Flow, 10E6 cu.ft/hr	7.387	2.351	1.178	0.640	0.385	0.234	0.200
Flow Area, sq.in. (150 ft/sec.)	1969.9	626.9	314.1	170.7	102.7	62.4	53.3
Delivery Pipe Internal Diameter, in.							
1 line at	50.00	28.30	20.00	14.75	11.40	8.91	8.24
2 lines at	35.40	19.98	14.14	10.42	8.09		
4 lines at	25.00	14.13	10.00				
8 lines at	17.70	10.00					

Table 1–6 Turbine Extraction Requirements

Removing the steam from the steam path can present a different type of problem. When removing the steam through an inner casing, it is necessary to ensure this does not adversely affect the geometry of the inner and outer casing, particularly when they move relative to each other during periods of differential expansion.

In determining the minimum axial gap required to remove the steam through the inner casing or between blade carriers, it is necessary to be aware of the inner diameter of the casing at the point of extraction. Consider Figure 1–34, showing the arrangement of stages in a portion of a reheat section. The outer sidewall of the diaphragms has been extended at a diameter *Do* to produced a gap *Ge* for removing the steam to chamber *Ch*. This chamber will be formed by the location of the inner casing portions one and two. From this chamber, the steam is removed through the outer shell through lines connected to the lower half and piped to the heater. To maintain the correct velocity, the extraction area $\pi.Do.Ge$ must be of sufficient size. This size can only be adjusted in terms of setting the gap *Ge*.

Similarly, the low-pressure section shown as Figure 1-35 shows the arrangement of a typical fabricated low-pressure casing, where three extraction chambers have been formed to allow the steam to be collected and removed to the feed heaters. In this arrangement the extractions are symmetrical.

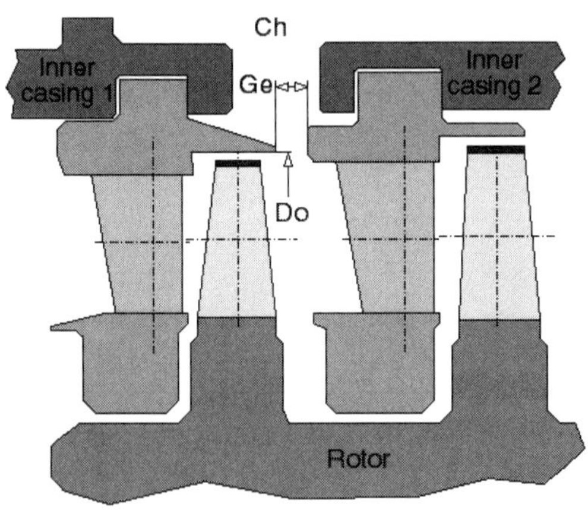

Fig. 1–34 The Arrangement of the Inner Casings or Blade Carriers to Permit the Removal of Feed Heating Steam

Fig. 1–35 Symmetrical Extractions from a Double Flow Low-Pressure Section

FLOW SPLITTING AND STEAM EXTRACTION

As the steam expands through the steam path, its specific volume and, therefore, volumetric flow increase. It is necessary to limit the axial flow velocities within the blades to levels that allow energy to be extracted as efficiently as possible. Therefore, even if mechanical constraints did not exist, there would eventually be a need to divide or split the flow into parallel paths to maintain acceptable levels of efficiency. There is also the need to remove steam from the expansion passages for regenerative feed water heating. These two requirements can have a significant effect on the form of the total steam path and the manner in which it is arranged.

Flow splitting is undertaken at the end of any expansion, where to continue in a single flow arrangement the radial height of the rotating blades would cause stress levels to exceed reasonable values. A suitable location, and one where blade heights are increasing at a rapid rate, is the end point in an intermediate or reheat section expansion. In many designs, the reheat expansion end point is selected to coincide with a required feed heater extraction point, and at that point the flow is divided into a suitable number of low-pressure sections—selected to be able to accommodate a suitable last stage blade configuration.

The optimum points in the turbine for extracting steam for regenerative feed water heating are determined in terms of the thermal ramp rate for feed water heating. The extraction quantities are determined on the basis of the amount of steam required to heat the feed water to the saturation temperature of the extracted steam at the heater inlet minus the terminal temperature difference designed into the heater. The actual and practical extraction points must be coincident with a stage end point, which places some limitation on the overall thermal gradient but is not as severe as might be expected. The reaction unit, in the high and reheat sections has a greater number of stages of smaller enthalpy drop. Therefore, it offers a better range for selecting extraction points and achieving a smoother gradient.

However, both impulse and reaction units can have their extraction steam conditions manipulated to a degree by minor modifications to the stage diameters that can be changed, thereby changing the stage enthalpy drop and the temperature at the stage end point. This can be done without compromising the performance of the stage—mechanically or thermodynamically.

Steam is normally removed from the steam path to undertake regenerative feed heating of the water being returned to the boiler. This water is condensate that was removed from the condenser.

Double-flow, high-pressure sections

A high-pressure section is designed to be double flowed only when the stresses due to centrifugal and bending effect in the rotating components are beyond the capability of the material to carry them safely. In fossil applications, it is uncommon to use double flow sections because the units can normally be designed to utilize the steam in a single-flow section and remain within acceptable levels of stress even at the highest temperatures.

In nuclear application however, for any unit sized above about 500,000 kW, double flow is almost always used. This double-flow arrangement is necessary because of the combination of low steam conditions (low-pressure and high specific volume when compared to the fossil unit) and the large flow quantities required to achieve design output. It is also normal for turbines to drive a four-pole generator, at 1,800 rpm, allowing larger diameters and a large axial area in the blade rows. For 50-cycle applications, the two-pole unit at 3000 rpm can often be utilized at larger ratings.

In the nuclear application, it is common for partially expanded steam to be extracted from the steam path either for feed heating, for the first stage of reheating in the two-stage reheat design, or both. A section through a double-flow nuclear high-pressure section is shown in Figure 1–36. The section shown has a symmetrical steam path.

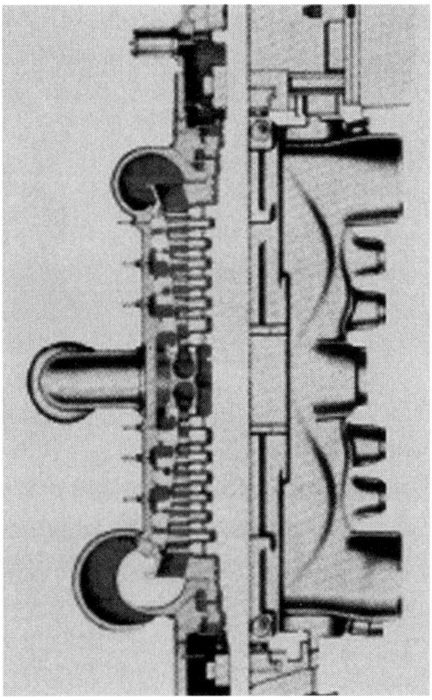

Fig. 1–36 A Double Flow Nuclear High-Pressure Section

Double-flow control stages in the high-pressure section

If a high-pressure section is double flowed, then the control stage will be of a double flow design with half the total flow passing through each of the two rows. However, there are some designs of single-flow fossil units where the control-stage blade loading is high enough that it becomes advisable to double flow this inlet or control stage rotating blades to keep the blade stresses within acceptable levels. This requirement is associated with the dynamic loading introduced by the partial arc admission effect. In a throttle-controlled unit this is not necessary.

Figure 1–37 is a double-flow control stage from a 530,000 kW unit with partial admission nozzle control. In this design, the inlet flow passes through parallel first stages and rejoins to flow through the remaining stages in the high-pressure section.

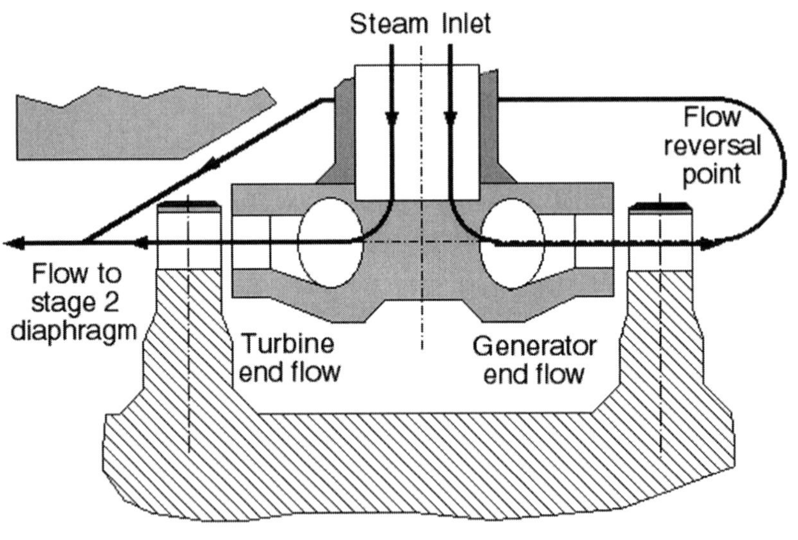

Fig. 1–37 A Split Control Stage

Extraction of partially expanded steam from the high-pressure section

If the cycle is designed to extract steam from the high-pressure section before it has completed its expansion, then the steam must be removed and passed to the heater before the main flow is removed from the turbine and returned to the reheater. This arrangement is called a heater above reheat point (HARP). Such an arrangement can—depending upon the high-pressure casing design and the possibility of flow reversal points—require that steam be extracted through a double-casing arrangement. The extraction of steam through a high-pressure inner casing can present certain levels of difficulty for the designer depending upon the details of casing and stationary blade support. These difficulties are dependent upon the need to remove steam through both an inner and outer casing. If the inner casing consists of several blade carriers, each located from the outer casing, then this difficulty is removed. An alternate solution is to remove the steam at a reversal point in the high-pressure expansion.

If the high-pressure section has a flow reversal, as shown in Figure 1–38, where flow is reversed after partial expansion, then this does provide certain advantages for the designer, including:

- It reverses the direction of the axial thrust, so the size and normal duty of the thrust block can be reduced.
- It reduces the temperature gradient across the inner casing, which reduces the thermal stresses induced during start-up, shut-down, and thermal transients.
- It reduces the pressure differential across the casing portions.
- It provides a suitable point in the expansion for steam removal for regenerative feed heating or some other function.

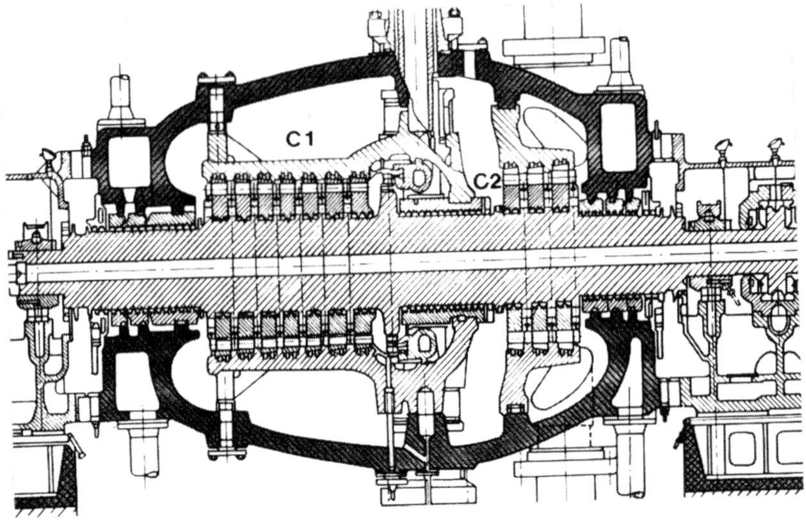

Fig. 1–38 A Reverse-Flow Design

Considerations concerning feed heating extraction. There are two determining factors when extracting steam for regenerative feed heating. First, determining a thermal gradient that will place a similar duty (temperature rise) on each feed water heater in the train. Second, for the top heater to achieve an FFWT consistent with that required for the cycle while allowing only the minimum amount of heat to be added in the boiler superheater section.

Extraction from high-pressure/reheat double-flow sections

If steam is to be extracted from a high-pressure or reheat section, the specific volume is normally sufficiently small that complexities are not introduced by the volumetric flow involved. However, in the case of double-flow reheat units, such as those used in the larger fossil designs and the high-pressure sections of nuclear units, there are advantages to extracting steam from both flows. This extraction from both ends is undertaken to maintain nominally identical steam paths in both flows. In such a design, these two flows of the steam path are of different hand, but the blading is of the same height and in all other respects identical. This arrangement can be used to allow a single blade design and to retain a balance between the thrust developed in both flows. The blading may be of opposite hand but otherwise identical. However, the sealing system may be different to allow for differences in the differential expansion from one flow to the other.

Nuclear high-pressure sections normally have a single feed heating extraction to the top feed heater at the exhaust (setting the FFWT) and in a two-stage reheat cycle will also remove steam for the first stage of reheating. Therefore, there are four extractions required in the steam path but they provide steam at only two extraction pressures. Therefore, the flows remain identical.

In the case of a double-flow fossil reheat section, there is no requirement for reheating steam but there can be as many as three extractions for regenerative feed heating. Therefore, other considerations that affect the extraction pattern are introduced.

Figure 1–39 shows a double-flow reheat section with steam removed for use in three feed heaters *A, B,* and *C.* This requires evaluation of extraction options and the configuration of the reheat section itself.

In Figure 1–39 the steam is extracted symmetrically, at pressures *Pa, Pb,* and *Pc* and flows through common headers to the three heaters *A, B,* and *C.* With this arrangement, because the same quantities are extracted from both flows at each pressure, the blading is identical, and the steam flow quantity through the two flows *T* and *G* is identical. With this design, the axial thrust developed on the blades rows is equal, and the theoretical net thrust of the two flows is balanced. The flow quantities from the turbine to the heaters would be *Qa, Qb,* and *Qc.* The quantities extracted from each end at each pressure would be the same.

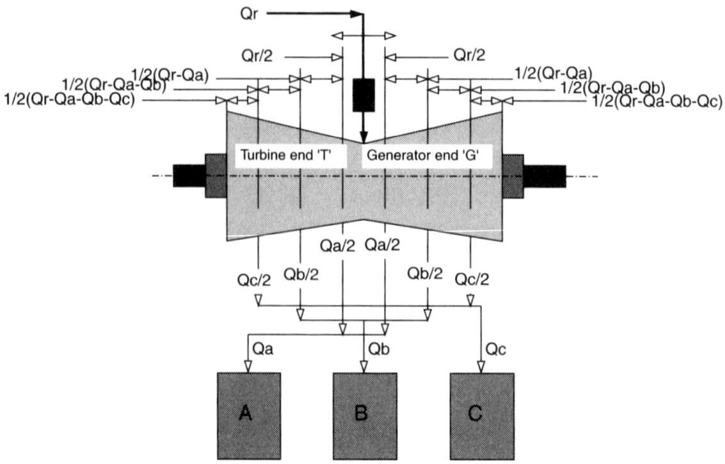

Fig. 1–39 Symmetric Extraction from a Double-Flow Section

In Figure 1–40 the flows extracted to the two top heaters *A* and *B* are from different ends of the unit, with extraction quantity *Qa* to heater *A* coming from the generator end only, and extraction quantity *Qb* to heater *B* from the turbine end only. Therefore, only the blading up to the extraction pressure *Pa* is identical.

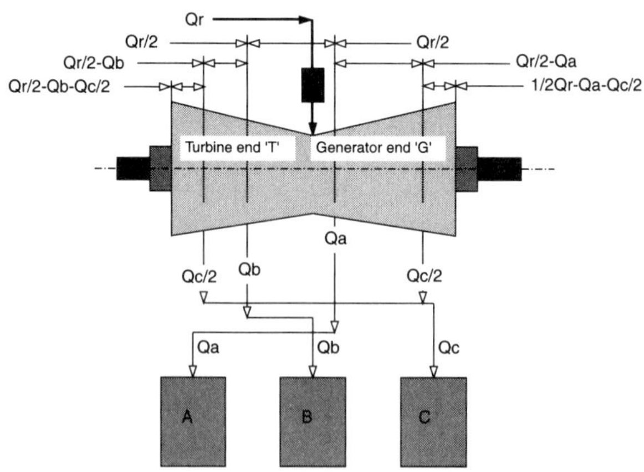

Fig. 1–40 Non-Symmetric Extraction from a Double-Flow Section

In an actual design, if it is required that the steam quantities through the last stage blades of the section are the same, then the initial quantities into the two flows *T* and *G* must be adjusted so that after extraction of quantities *Qa* and *Qb* the remaining flow after heater *B* extraction to exhaust is identical. At a pressure *Pc* steam quantity *Qc* is extracted symmetrically to heater *C*, quantities *Qc/2* being extracted from each end.

If the extraction quantities *Qa* and *Qb* at pressures *Pa* and *Pb* are different, then they will have caused an imbalance in the flow. Also the quantities discharging to the low-pressure sections from each flow will be different. With this extraction arrangement, the axial thrust developed in the two flows will be unbalanced and require the unbalanced portion be carried by the thrust bearing. This is not a major consideration, but the operator should be aware of this.

Should it be necessary to isolate heater *A* or *B*, then there is an adjustment of flows through the blade system, and some small modifications of the pressure distributions in the blade rows.

Because of the differences that exist in the extraction quantities required for each heater, there are various philosophies used to determine the quantity of steam entering the two flow sections *T* and *G* and also how the steam is directed upon removal from this double-flow section. The overall arrangement for directing and distributing flow is also influenced by the number of low-pressure expansions used and the form of the piping carrying the steam to these low-pressure sections.

Important consideration in selecting extraction configurations throughout the unit are the steam loading placed on the last stage blades and the discharge velocity loss that occurs at their exhaust. To minimize this velocity loss, it is necessary to have equal (or near equal) flows through each exhaust in the unit. The steam loading developed on the blades is in direct proportion to the quantity of steam flowing through the row, which represents another reason for equalizing this last stage flow. There is often a need to adjust the flow through the various sections before the last stage so the discharge velocity remains the same at all exhausts. To equalize this flow, there should—or must be—some modification made to the quantity of steam entering the first stages of the units.

To examine the possible arrangement of flow directions and splitting, consider some of the arrangements that can be made in flow distribution. To do this, examine the requirements of both four- and six-flow expansion arranged in multi double-flow sections with these sections receiving steam flow from double intermediate (reheat) expansions ahead of them.

To achieve equal or near equal flow through the last stage (L-0) blades, the possible arrangement include:

- As shown in Figure 1–41a, steam discharges from the double-flow reheat section in two quantities *Qd1* and *Qd2,* with each reheat exhaust line feeding one double-flow, low-pressure section. Equal flow through the exhaust stages, with symmetrical extraction from *LP1* and *LP2,* can be achieved only if the extractions from the reheat are symmetrical as shown in Figure 1–41a. That is, *Qd1* must equal *Qd2.* In fact, the differences between *Qd1* and *Qd2* could be sufficiently small that the values of *Qe1* and *Qe2* are sufficiently small enough that this physical arrangement of the steam path can be accommodated.

- If the discharge from the double-flow reheat section is sent to a common header (crossover pipe), the requirement of symmetrical extractions in the reheat section does not exist. In this situation, as shown in Figure 1–41b with symmetrical extractions from the low-pressure section *LP1* and *LP2,* the values of *Qe* are identical in all four exhausts.

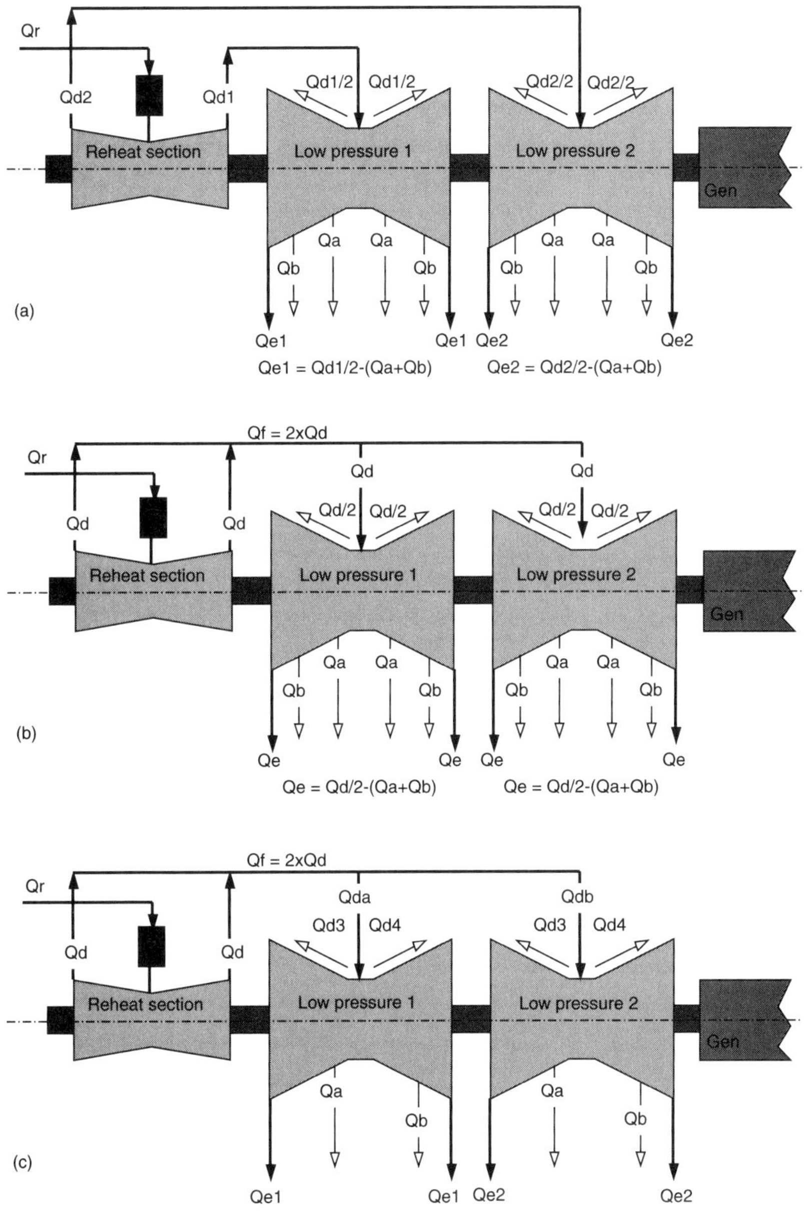

Fig. 1–41 Possible Arrangements for Using Steam Extracted from a Four-Flow Low-Pressure Design

- In the case of non-symmetric extractions in the low-pressure sections, the only means of ensuring equality of last stage flow, as shown in Figure 1–41c, is to adjust the quantities of steam entering the low-pressure sections. That is, adjust the inlet flows in both sections so that *Qe1* is equal to *Qe2*. This done by adjustment of *Qd3* and *Qd4* so that:

$$(Qd3 - Qa = Qe1) = (Qd4 - Qb = Qe2)$$

This flow adjustment is achieved by modification of the discharge areas of the first row stationary blades in the low-pressure sections. The flow division in these first rows can be considered to provide an opportunity division that occurs in the ratio of the area through which the flow must pass.

Low-pressure extractions from multi-flow sections

Multi-flow, low-pressure sections can be arranged to have two, four, or six exhausts in a tandem arrangement. Cross compound arrangements have been manufactured with as many as eight exhausts with two double-flow sections on each of two lines. Because of the large energy range of these low-pressure sections, and the temperature variation that occurs, there are normally three or four extraction points for the heater train from the low-pressure sections. The design of a low-pressure section with large exhaust flows is a complex process taking a number of years to design, undertake model and field tests to prove the design sufficiently, and ready the design for offer within the market. For this reason, it is required that any design is suited for multi-section application. The most critical components in the low-pressure sections are the *L-0, L-1,* and in some applications, the *L-2* stage rotating blades. For this reason, these are standard components that the designer is not prepared to modify unless there is some compelling mechanical or structural reason for doing so. It is also normal at the lower pressures regions that the volumetric flow is so large feed heating quantities cannot be accommodated by removal from one section.

For those blade rows ahead of these critical components minor, modifications can be considered. However, if such changes are made, they are not normally on a contact specific basis but rather to represent a change that provides an improvement in either efficiency or reliability of the unit and that in the future will be offered for other units within the fleet of those designs.

For these reasons—maintaining interchangeability and standard designs—the low-pressure section design process becomes more complex. The considerations that relate to the selection of the different extraction arrangements and section designs relate to the possible interchangeability of rotors and the ability to carry a common spare that will fit into any of several sections within a unit. From both a manufacturer's and an operator's perspective, there is an incentive to make the low-pressure sections duplicate designs so a considerable level of interchangeability can exist. This is of particular interest to operators in multi-unit stations, where for important base load installations complete spare rotors can be carried to minimize outage time should any form of mechanical damage occur that prevents a rotor from being returned to service. The cost of carrying a single spare rotor can often, in terms of reducing unit forced outage rates, offset any costs related to the purchase of this replacement element.

The question of rotor interchangeability and extraction configuration is therefore one that receives considerable attention when a prototype section is being designed. Since a double-flow low-pressure section can be used in a single- or multi-section configuration, it is necessary to preserve interchangeability so each of the double-flow sections should have totally interchangeable rotors. Major considerations in the determination of low-pressure design are

- The lowest pressure extraction will normally, because of large volumetric flows, (see Table 1–6) require steam be removed from both expansions (the turbine and generator end).
- If there are three extractions from a double-flow section, it is normal for the other two higher pressure extractions to be arranged to remove steam from either end.
- If there are four extractions, it is possible to remove steam for the lowest two pressure heaters from both ends and for the two highest pressure heaters to be arranged to remove steam from alternate ends.
- Non-symmetric extractions to the highest pressure heaters will cause a thrust imbalance that must be carried by the thrust bearing.
- There is a need to remove moisture from these units, so there will is a need to drain each stage in the moisture region where steam is not to be extracted. It is necessary to cascade drains, sometimes for several stages, with different drain designs in the two flows.
- The positioning of moisture collection grooves is different in each of the stationary sections.
- The differential expansion is different in each flow depending on the location of the thrust block. Therefore, these sections require different axial clearances between the stationary and rotating blade rows. It is necessary to design the cold settings in each section to suit the differential expansion the section will experience in operation.

There is, therefore, a fundamental decision required of the designer in the case of a new design of double-flow low-pressure sections regarding whether to make the sections symmetric or non-symmetric in terms of extraction arrangement.

Possible low-pressure extraction configurations

There are two possible extraction patterns from double flow low-pressure sections that are required to supply steam to four different pressure heaters. These are

1. **Sections with symmetrical extractions.** The same quantity of steam is extracted from each end for each extraction pressure and the same quantity of steam is removed from both ends at the same pressures.

2. **Sections with non-symmetrical extractions.** These sections are arranged to remove feed heating steam from both ends but with different quantities and pressures at each expansion. The exception to this is normally the lowest pressure. Due to volumetric flow requirements, steam is removed from both flows at the same pressure.

These two concepts for extraction are shown schematically in Figure 1–42. The symmetric design is shown Figure 1–42a, where four extraction locations at pressures *Pa, Pb, Pc,* and *Pd,* are shown. *Pd* is the extraction at entry to the exhaust stage. Figure 1–42b and c shows other arrangements where steam is extracted but only at three pressures—*Pa, Pc,* and *Pd* in (b) and pressures *Pb, Pc,* and *Pd* in (c). In Figure 1–42d, the steam is extracted at all four pressures, but pressures *Pa* and *Pb* use only one extraction per end. These extraction arrangements can then be used in different combinations for multi-flow low-pressure sections.

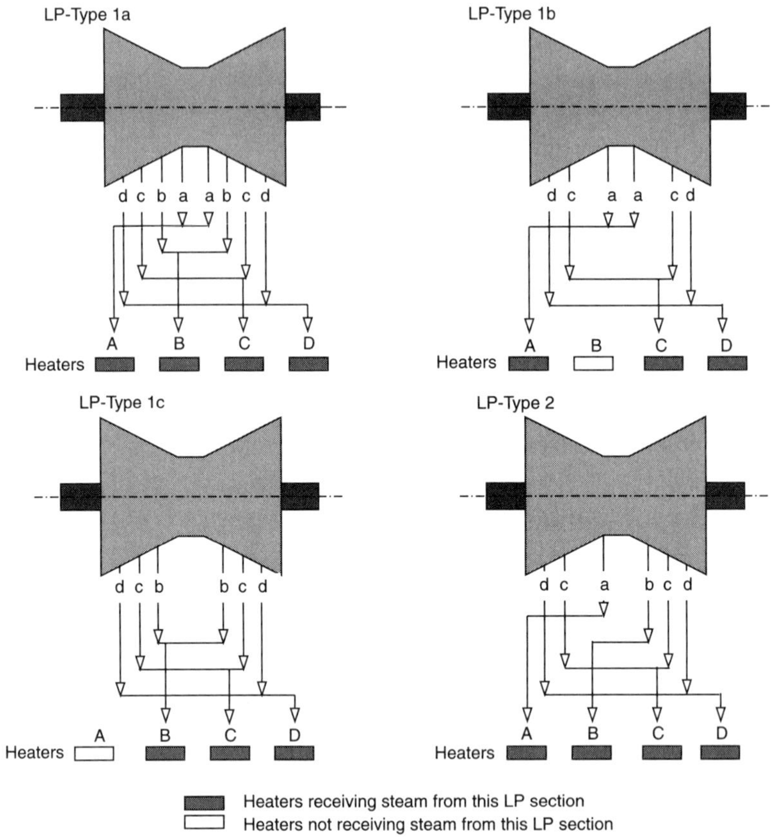

Fig. 1–42 Various Configurations Both Symmetrical and Non-Symmetrical Double-Flow Low-Pressure Sections

Four-flow units. Four-flow arrangements can be either symmetric or non-symmetric depending on the design philosophy used and the need for interchangeability. Figure 1–43a shows an arrangement of the rotors in which the extractions are a symmetric arrangement in each rotor, but the rotor would not be interchangeable. Figure 1–43b is a similar four-heater arrangement using non-symmetric extraction, but the rotors are identical and the rotors would therefore be interchangeable.

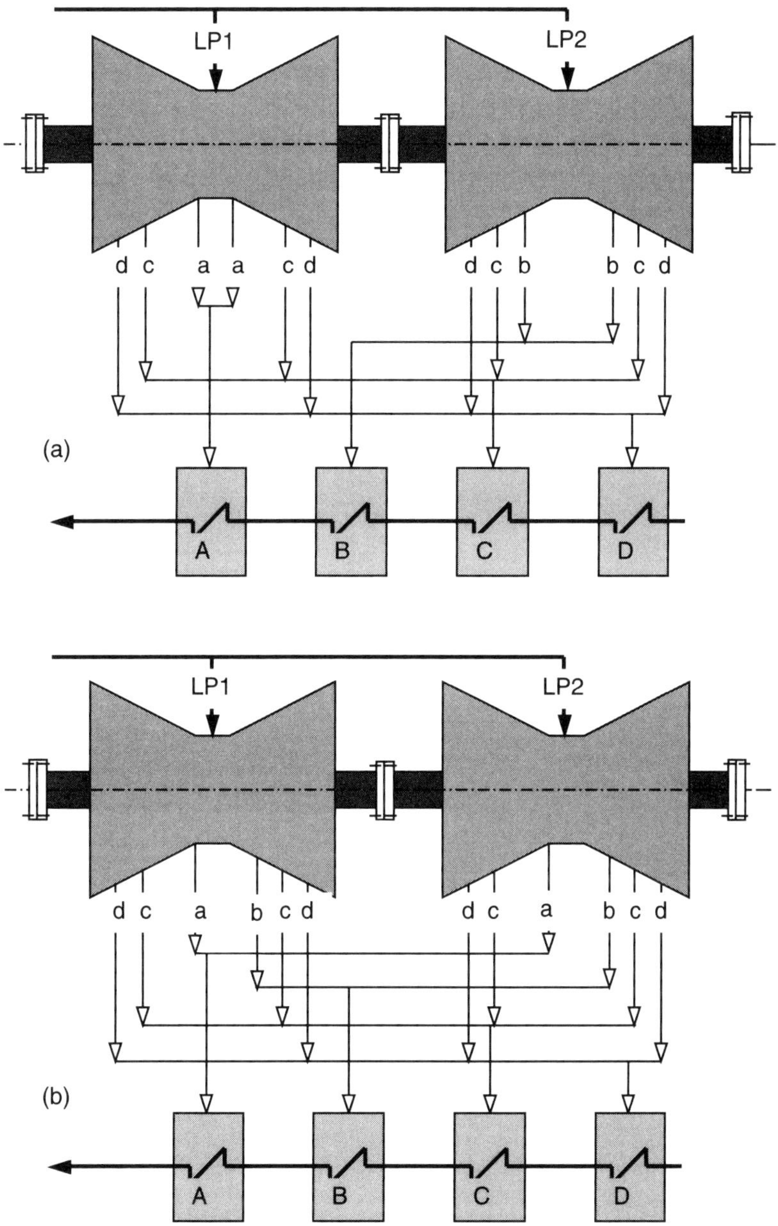

Fig. 1–43 Extraction Patterns

Six flow units. Figure 1–44 shows two possible arrangements for a six flow unit.

Figure 1–44a shows a design in which the extraction from the rotors is a non-symmetric arrangement, with pressure extractions *Pc* and *Pd* removed from different ends of the unit. However, these rotors can be made interchangeable.

The extraction arrangement is:	LP1 Pressures	Pa	Pb	Pc	Pd
	LP2	Pa	Pb	Pc	Pd
	LP3	Pa	Pb	Pc	Pd

Figure 1–44b is another arrangement with individual rotors that are symmetric from flow to flow. However, in this arrangement rotors *LP1, LP2,* and *LP3* would not be interchangeable. The extraction pressures from the three sections follow.

The extraction arrangement is:	LP1 Pressures	Pa	-	Pc	Pd
	LP2	Pa	Pb	Pc	-
	LP3	Pa	Pb	-	Pd

In any design with a low-pressure section to be defined and that is to become a standard for the turbine builder, the designer is left to decide whether a non-symmetric arrangement such as shown in Figure 1–44a is to be used. In this design, extraction steam is removed from the low-pressure sections at three locations on either end, supplying steam to four feed water heaters. For example, the extraction point on each turbine end *T* goes to heaters *D, C,* and *A,* while steam extracted from the generator end *G* is removed to heaters *D, C,* and *B.* In each section of this design, the three low-pressure sections are identical, but have different steam paths at the turbine and generator ends.

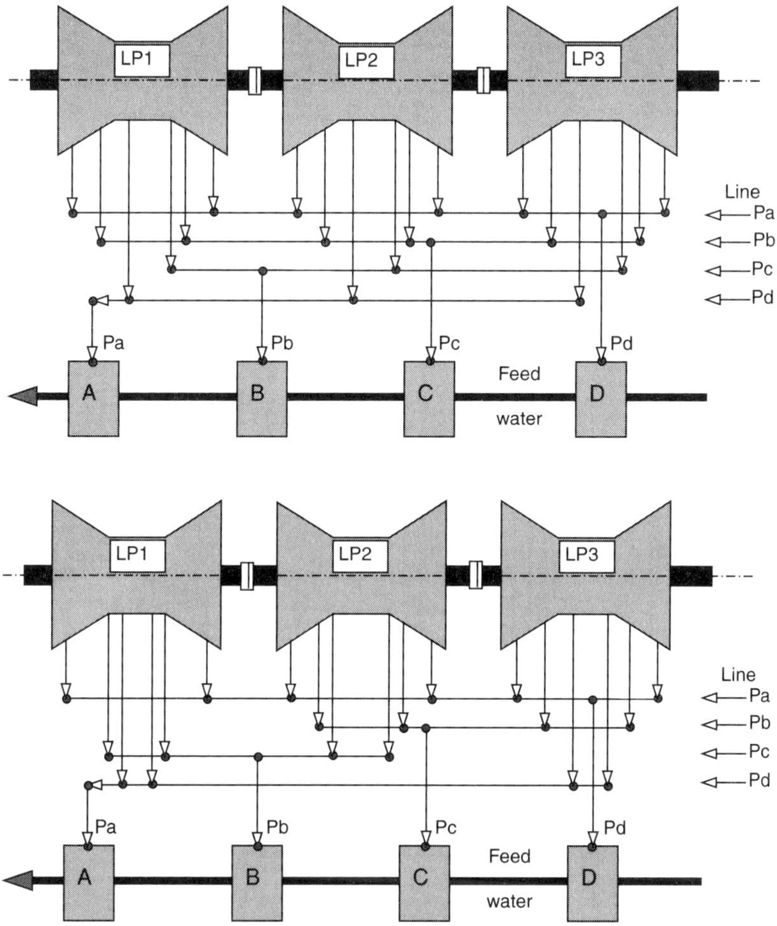

Fig. 1–44 Alternate Steam Extractions Patterns from a 6-Flow Low-Pressure Design

These three low-pressure rotors of Figure 1–44a can be interchangeable if the axial clearances between the blade rows are retained at the same values in all three. Alternately axial clearance differences can be minimized by adjustment of the stationary blade row axial position setting. Under these circumstances, only one rotor design and one spare would be required for all three sections. The axial gap between stationary and rotating blades may need to be set at a constant in all three sections due to differential expansion requirements and the need to preserve the interchangeability absolutely. However, the axial setting of the diaphragms or blade carriers can also be set so that the clearances in each of the three sections are optimum.

If the axial gaps are set equal in all three sections, there is a small and difficult-to-quantify efficiency loss associated with larger-than-necessary axial gaps. This loss is small compared to the advantages of interchangeable rotors, particularly in multi unit stations. For gaps smaller than optimum, the blading losses are greater than for gaps that are larger than the optimum.

With identical rotors, because the differential expansion in all three sections is different and is related to their axial distance from the thrust block, the axial clearance requirements between the stationary and rotating blades in each of the three sections must be considered and evaluated separately. To achieve and maintain interchangeability, this axial spacing between the stationary and rotating blades must be adjusted so any rotor can be placed into a casing without fear of interference during operation due to the local differential expansion. It is possible to optimize the axial spacing by adjusting the axial placement of the stationary blade row. However, it is normal to accept some level of compromise in the axial clearances to maximize interchangeability.

A major consideration in the design of low-pressure sections is that it is difficult because of large axial movements in the low-pressure sections to effectively utilize axial seals on the blade rows. Normally in such a design, all blade tip seals will need to be radial.

An alternate arrangement to that shown as Figure 1–44a is that shown as Figure 1–44b. This is a design in which the rotors are symmetrical from one flow to the other, but one that will not achieve identical or interchangeable rotors and does not achieve flow symmetry.

In this design, the bottom heater *D* is supplied with steam from each of the six flows, which may be convenient and necessary because of the volumetric flow involved and the need to maintain steam velocities at acceptable values. However, each of the remaining three heaters are supplied with steam, from two of the three double-flow low-pressure sections. For heater *C*, steam is removed from low-pressure sections two and three. For heater *B*, steam is removed from low-pressure sections one and two and for heater *A* from low-pressure sections one and three. With this design, there will be a difference in blade heights except for the last two stages where steam is removed ahead of the *L-1* stationary blade row and before the *L-0* stationary blade row. Therefore, it is possible to arrange elements so the steam path and blade requirements are identical in these two stages but of different hands. In this case, there will be a net axial thrust of zero developed on all three rotors with each section achieving a balance between the *T* and *G* flows. There is, therefore, no load developed on the thrust block. Also because each section is a discrete design the efficiency can be optimized.

With this extraction arrangement, the steam quantity flowing to the turbine and generator end must normally be adjusted to ensure the correct quantity flows into each of the halves. Then after steam extraction differences, the flows through the last two stages of each expansion are the same. That is, the exhaust velocity through the last stage blades and the steam quantity through the exhaust row on all six flows are the same.

Important considerations for these two designs are those relating to the repair of the first-stage nozzle by weld rebuild. The areas of these first stage nozzle boxes must be rebuilt to ensure the steam quantity to the three flows are adjusted correctly. If this is not done, there can be uneven steam flow to some sections, possibly introducing excessive loading on the latter stage rotating blade rows, causing an excess velocity, a higher leaving loss, and higher than designed operating stresses.

The designer is left to determine whether a symmetric or non-symmetric arrangement will provide the most flexibility for future offerings.

The design decision will normally be made on the basis that the newly designed section can be used in a two, four, six, or eight exhaust configuration. There are some designs that make it relatively easy—by blocking and not using certain extraction pockets—to make the sections either symmetric or non-symmetric and then modify the diaphragm axial spacing to preserve the axial clearances at or near optimum values. Designers can also make relatively minor changes to the fabrication design of the low-pressure hood carrying the diaphragms to eliminate extraction pockets. However, this does represent different designs and once equipment is installed it is difficult or impossible to make changes to the diaphragm axial placement.

REFERENCES

Baily, F.G., K.C. Cotton, and R.C. Spencer. "Predicting the Performance of Large Steam Turbine-Generators Operating with Saturated and Low Superheat Steam Conditions." Presented at the 29th Annual Meeting of the American Power Conference, April 1967 Chicago, IL.

Keller, A. and J.E. Downs. "Effect of Exhaust Pressure on the Economy of Condensing Turbines:" Presented to the Power Division of the American Society of Mechanical Engineers (ASME) Power and Hydraulics Division, Los Angeles, CA. July 1953.

Salisbury, K. J. *Steam Turbines and Their Cycles.* Huntington, N.Y.: Robert E. Kreiger Publishing Company, 1974.

Spencer, R.C., Cotton K.C., and C.N. Cannon. "A Method for Predicting the Performance of Steam Turbine Generators. 165,000 kW and Larger." ASME Paper 62-WA-209, Annual Winter Meeting, New York, N.Y., 1962.

Spencer, R.C. and J.A. Booth. "Heat Rate Performance of Nuclear Steam Turbine Generators." Presented at the 30th Annual Meeting of the American Power Conference, April 1968, Chicago, IL.

Chapter 2

The Turbine Steam Path Mechanical Design Process

INTRODUCTION

The mechanical design of the turbine steam path begins after completion of the selection of the cycle equipment, its arrangement, the determination of the number of stages, and the thermal sizing of the blade rows. The thermal design provides information on the number of stages, the blade geometries, and the variation of thermal conditions throughout the steam path. At completion of the thermal process, there can still be a number of thermal variables that are to be evaluated by the mechanical process, but essentially the scope of the unit and its general arrangement has been established.

Normally, before the heat balance has been prepared, the thermal design engineer and the mechanical designer will have determined the basic arrangement, including the number of flows at each pressure range, the points at which steam will be extracted for regenerative feed heating, and any other extractions which may be required.

Therefore, the requirements of the mechanical process will, in general, have been defined, if only in broad scope. What the mechanical process is to do, is size the various components so thermal requirements are met and stress levels and natural frequencies are within acceptable ranges that will not endanger the structural integrity of the unit.

THE MECHANICAL DESIGN PROCESS

The mechanical design process accepts the detailed data developed by the thermal design engineer. Then, to complete the mechanical design of the unit, there is a certain amount of data that must be provided from the thermal sizing calculations, and the requirements specified in the final design specification must not compromise these requirements. To do so could compromise the thermal performance of the unit. The thermal calculations provide the stationary and rotating blade requirements, the number of stages, and details concerning the general arrangement of the unit sections.

These thermal calculations also provide boundaries that must be achieved to meet output requirements and within which the mechanical design must be achieved. These calculations also define the steam flow quantity and the steam conditions to be achieved at various locations within the unit and cycle. They also establish blade heights, mean diameters, and vane angles requirements. The thermal definition calculations provide information that can normally be achieved by using a number of existing design components, from which the mechanical designer can select those most suited to the particular application—those capable of achieving both efficiency and reliability goals.

However, the mechanical design and the design specification it produces are not limited to meeting the requirements of the thermal specification. It must also specify a number of other requirements.

The tolerances within which components are to be manufactured. These tolerances define the dimensional boundaries for each component. While such information may be placed on manufacturing drawings, it may also be provided as an engineering specification with the manufacturing drawing prepared by a manufacturing engineering group.

In the majority of steam turbine manufacturing companies' individual drawings do not normally include dimensional tolerances. Rather, these exist as works standard sheets, which are published as those quality requirements to be observed throughout the manufacturing departments.

The requirements of surface finish. There are certain components where the surface finish is critical from both an efficiency and reliability perspective. Where such a finish is required, this can exist either as detail on the manufacturing drawing or as works standard sheets.

Material of construction. There are a variety of materials used for the construction of a steam turbine. Many of these materials have been developed to make them suitable for long-term application at high pressures and temperatures. There are a number of components, such as rotating blades, threaded components, and seal systems, where a variety of materials are available for use at different locations throughout the steam path. The mechanical designer must select between these alternates and establish what particular material is to be used for any specific location within the unit.

Special processes. Certain aspects of design require the application of processes, which at completion cannot be gauged for adequacy. In such circumstances, the design engineer will define the requirements of the process in terms of application, possibly the qualifications of the person to undertake the work, and the equipment needed. This could be provided as a process specification. Such process specifications and other details exist as works standards.

Special functional tests to be conducted before use. These tests are conducted to ensure that certain components meet design specifications or that an assembly is ready for installation at a supplier's plant. These can include overspeed testing, hydrostatic testing, and any other major functional test, many of which a purchaser might wish to witness.

Non-destructive tests. Such tests are specified together with the acceptance level for any defects that are determined to exist.

Information required from the thermal design

The thermodynamic design output provides certain information to the mechanical designer, and does in fact define the boundaries within which the mechanical design must be accomplished. Some of these thermodynamic properties are critical and the mechanical designer has no authority to modify or change them in any way. To do so would cause a deterioration in the performance potential of the unit. Other properties are based on the mechanical design assumptions made by the thermodynamic designer. These represent boundaries or other properties that can be modified without any significant impact upon the performance of the final design, and the mechanical designer will evaluate these and modify them as necessary.

The major thermal output quantities or constraints are as follows:

The steam mass flow rate *m*. This quantity defines the steam flow for each stage. This flow quantity will normally indicate the quantity of steam passing through the stationary and rotating blade rows. This flow quantity can also include information on calculated or assumed quantities of leakage steam that is bypassing the blade elements in both the stationary and rotating blade rows and at other points where leakage can occur. The mechanical designer must confirm these quantities after some determination has been made of the optimum seal arrangement. This information includes the number of series seals, the diameter at which they are effective, and the clearance in either the radial or axial direction—whichever is appropriate for the stage configuration.

The thermodynamic calculations also provide information on steam quantity change. These represent the amounts of steam added to or subtracted from the main steam flow. Such increases or reductions would be due to extraction from the steam path for regenerative feed heating or process application, the return to the main flow of steam from a process, or leakage return from some other portion of the cycle.

The steam conditions. These properties include a definition of the value of the enthalpy H at inlet to and discharge from the cylinder and at each stage point. This portion of the thermodynamic data will also contain information on the steam pressure P and temperature T at each of these same positions. These quantities are sufficient that other qualities or properties of the steam, such as the specific volume Vs, together with physical properties required to allow a complete mechanical design and final analysis to be made. For flow in the saturated region, the moisture content and its variation throughout the expansion can be established.

Depending on the length of the blade vanes and the stage diameters, there could be a considerable variation of the steam properties along the radial length of the vanes of both the stationary and rotating stages. These variations can be sufficient so there is a need for them to be accounted for in the design of the vane profiles.

Characteristically, the Mollier diagram for a stage is presented for the mean (or reference) diameter. Such a Mollier diagram for a stage mean diameter is shown in Figure 2–1b. Here the stage has an inlet pressure of *Pi* at enthalpy *Hi* and an exhaust pressure *Pe* at enthalpy *He*. The isentropic energy at pressure *Pe* is *He*. There is a pressure between the stationary and rotating rows of *Pm* at enthalpy *Hm* (isentropic). With this pressure and energy distribution, the total enthalpy drop across the stage is divided into two portions *Hs* in the stationary and *Hr* in the rotating row. However, this pressure *Pm* represents the pressure between the stationary and rotating blades at one radial position only, the mean diameter, and there is a radial pressure gradient in this inter-row gap.

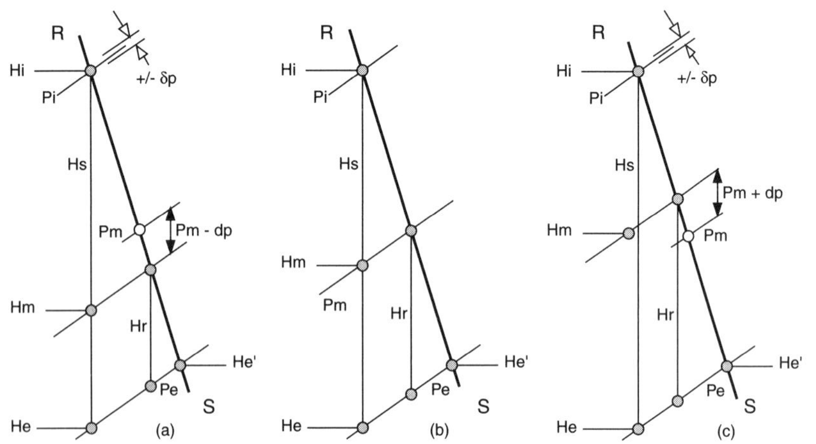

Fig. 2–1 The Mollier Diagram at Three Radial Positions

The steam, as it discharges from the stationary blades has a velocity with a radial flow component. The effect of the radial flow component in the steam velocity is to modify this inter-row pressure *Pm* along the radial length of the blade, increasing it toward the tip. The Mollier diagrams at the root section are shown as Figure 2–1a and the tip section is Figure 2–1c The major difference in these three diagrams is the variation of the inter-row pressure.

At the root section, because the steam has had an outward flow effect, the steam pressure has reduced by an amount *Pm-dp*, reducing the energy which is available across the rotating blade at this section, but increasing the stationary blade energy and the steam discharge velocity from the stationary row. Similarly, at the tip section, the pressure at the inter-row position has increased to *Pm+dp*, making more energy available across the rotating blade but less across the stationary row and reducing the discharge velocity from the stationary row.

There is also the possibility the pressure at entry to and discharge from the stage will not be completely constant along the radial positions. However, such variation is relatively small, dependent upon stage geometry, and can be represented on these diagrams by the amount $+/-\delta p$.

There is a pressure gradient from the root-to-tip section in the inter-row position, and the absolute pressure drop across the stationary blade decreasing towards the tip. Because the velocity at discharge from the stationary blade is a function of the thermal energy *Hs* across the row, there will be a continual variation (decrease) in the discharge velocity from the root-to-tip sections. In addition, the tangential component of the rotating blade velocity is a function of its diametral position. There is therefore a variation in the term *U*. Both these factors combine to modify the velocity triangles and the rotating blade geometrical requirements along the radial height of the vane.

In considering the stationary blades of any row, the variation in the stage thermodynamic parameters are sufficient, and the velocity of the steam entering the rotating blade row varies as a function of radial height. These variations of velocity will require the inlet portion of the rotating blade be designed to accommodate these changes in the relative velocity of the entering steam. Similarly there is a variation—with radial height—of the thermal energy *Hr* which is available for conversion in the rotating blade. This available energy variation together with that of the tangential velocity *U* will modify the rotating blade absolute discharge velocity and cause a change in the requirements of the discharge portion of the rotating blade along its radial length.

The thermodynamic specification should provide sufficient data on the required variation of steam conditions and physical properties such as velocity and absolute discharge angles along the length of the blade, so a suitable mechanical design can be produced. Depending upon the length of the blade and the thermodynamic design method, the thermodynamic specification may provide data at various stream-line positions.

The thermodynamic specification should also provide some data on the physical requirements of the stage. This will include a definition of the discharge area *Ad* required for each of the rows. This area must be achieved by the mechanical design, and the unit will eventually be manufactured to achieve this discharge area within close limits. This is necessary because this discharge area will establish the stage pressure and, therefore, enthalpy at the inter-stage points.

The thermodynamic design will also select the stage mean diameter *Dm*. This becomes a major design parameter that cannot be modified, and it is used to define the enthalpy drop on the stage required for maximum efficiency. Similarly, the stage discharge height *h* will be defined based on certain mechanical assumptions that must be confirmed. The mechanical design process will select and possibly design profiles to achieve the required throat at each radial position. The throat and its variation as a function of vane height are required to achieve the design discharge area.

When this thermodynamic information has been made available, the mechanical designer is in possession of sufficient information to investigate and evaluate the options available to him and to provide a suitable arrangement of series stages arranged to expand the steam. Each of the stages will consist of a stationary and rotating blade pair that are arranged to provide a controlled expansion of the steam in discrete pressure drops. The individual stage pressure drops are selected to optimize the expansion and energy conversion process. Expansion of the steam through these stages will develop as much rotational kinetic energy in the rotor as possible, and the mechanical design process will select components to minimize the fluid and mechanical losses which are induced in the individual stages.

The mechanical design requires a series of profiles designed to accept steam from the previous row and discharge it to the following blade row. The details of the stationary blade row will require a vane, an attachment to the stationary portion of the unit in the form of location within a diaphragm or directly into the turbine casing, and an inner seal arrangement where the blades are in close proximity to the rotor. The rotating blade row will need an acceptable vane, (either cylindrical or vortex), a root attachment for transference of the loads from the blade to the rotor, and possibly tie wires and coverbands.

For high-performance modern units, most manufacturers have available families of standard components with vane profiles of known and proven characteristics that can be used over various lengths and in various widths. In addition to the availability of standard profiles, manufacturers with advanced production techniques have available families of vane profiles that can be varied or manufactured at standard rates of taper or twist. The vanes, which these tapers define at various heights, establish a known rate of variation of section properties along their length.

In addition to these standard vane profiles, manufacturers have standard root form profiles that can be produced for various pitches and wedge angles to suit different rotor diameters and number of blades in the row.

Each manufacturer also has normal preferences that suit its basic design and various formats for tie wires and coverband forms and arrangements. The tie wire form used depends on the type of blade and the characteristics of the stage. The coverband reflects the preferred method of producing a seal and it is also dependent on the extent of differential expansion in the stage.

Vane profiles and other stage details are considered standard for any modern manufacturer and are normally produced through extensive research. Such components represent a product that is efficient and provides load carrying capabilities consistent with the stresses the elements are expected to experience during operation. When a vane is available at a standard rate of taper or vortex form, these profiles and their rate of variation are determined by detailed calculation and possibly research. This research is intended to define a profile that combines high efficiency with acceptable mechanical strength.

Mechanical considerations

The mechanical calculations performed in selecting and arranging the various components comprising the stages cannot compromise the thermal guarantees made in the heat balance. There are, under certain circumstances, compelling reasons to deviate from the design specified requirements, but such changes—if they are necessary—must be referred to the thermal design engineer for changes in efficiency predictions and to determine whether such a change will compromise the guaranteed efficiency to an unacceptable degree.

The design procedures used by any manufacturer are dependent in part upon the procedures established within his company. The stationary and rotating blade row general requirements are determined by the thermal design engineer. However, the investigation of stress levels and vibration characteristics of the stages is a responsibility of the mechanical design authority. A typical design process for the steam path is shown in Figure 2–2, where vanes are selected to meet thermal requirements. Then stage hardware such as coverbands and tie wires are selected and their sizes, grouping, configuration, and their adequacy are determined by mechanical analysis.

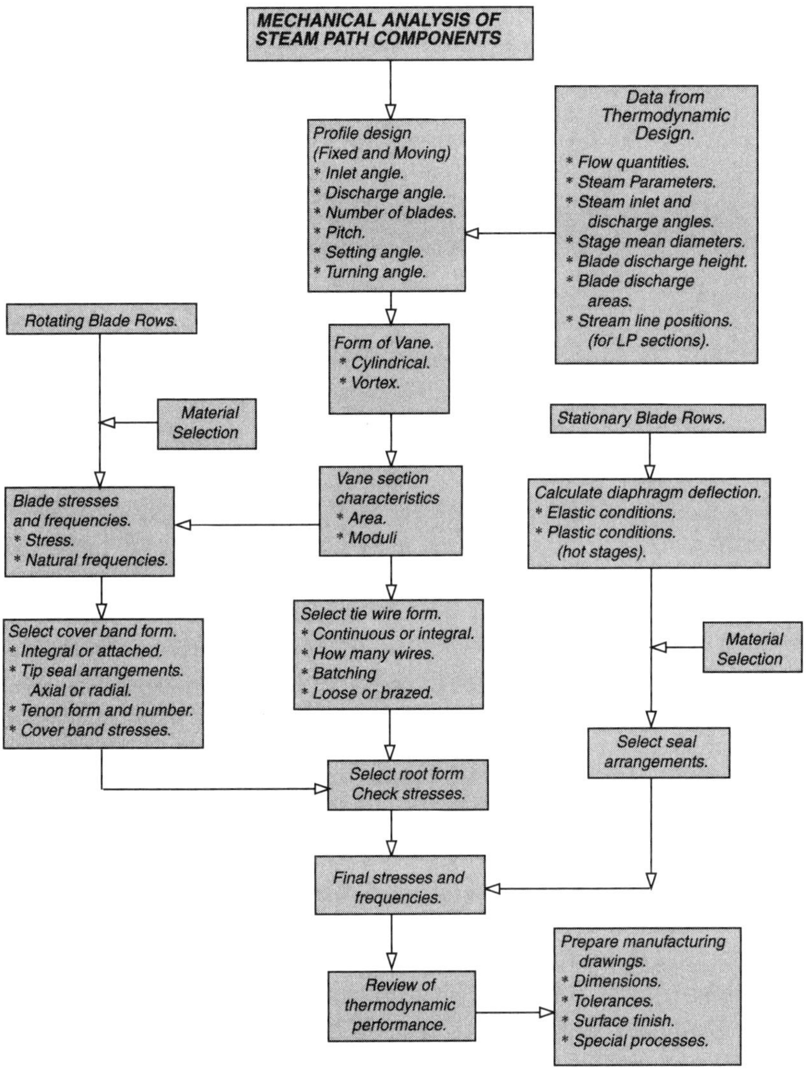

Fig. 2–2 The Mechanical Design Process for a Typical Stage (See Fig. 1–1)

Once the basic details of the steam path have been established by the thermal calculations, the following information can be established:

- The axial spread required for individual stages is determined from the requirements for rotor and wheel stress and the space between the wheels to accommodate the diaphragms.
- The span of the individual rotors, which require a knowledge of the individual stages and the location of the shaft-end sealing provisions, including the number of seal strips that can be accommodated.
- The diameters of the rotor at all axial positions including over the blade tips. In the case of a drum rotor, this will also require the diameter and general arrangement of the pressure balance piston.
- The axial space available for seal system placement and the number of series seal strips that can be used.
- The inside diameters of the casing, which then allows the casing form to be established.

When the steam path dimensions—including stage spreads and diameters—have been established, there are a number of supporting calculations that must be undertaken before final dimensions can be established. These supporting calculations include:

The determination of the rotor critical speeds. These are important in the cold uncoupled condition. This is a requirement so that a definition of the balancing and overspeed tests can be made. It is also necessary to determine the hot-coupled criticals, as this represents the normal operating condition.

To determine the shaft deflected form. It is necessary to determine this deflection in the cold condition so an alignment determination can be made. It is also necessary to establish whether there is any significant additional deflection so there is assurance that radial clearances will not be compromised during operation.

The diaphragm axial deflection under load. It is necessary to predict the amount of diaphragm axial deflection under the pressure differential that exists across it. For hot stages (higher than approximately 750°F), it is also necessary to consider plastic deformation, or creep, to be sure axial clearances between stationary and rotating components will not be consumed during the operating life of the unit. This operating life is considered to be 200,000 hours—equivalent to 30 years operation at a load factor of about 75 to 80%. In extreme cases it would also be necessary to ensure that creep deflection did not reach the tertiary phase where mechanical rupture would occur.

Short circuit torque. It is necessary to predict the shaft torque in the event that a serious short circuit condition occurred, where a phase-to-phase condition would be possible. It would be necessary to check the rotor torque at all locations to be sure this would not cause extensive damage.

Differential expansion. The extent of differential expansion under start-up, normal operation, and at shut-down conditions must be determined to ensure the axial clearances throughout the steam path are adequate. This will also help define where radial tip seals rather than axial are to be used.

Rotor and casing radial growth. The extent of radial growth when the unit is hot and at speed needs to be examined. It is also normal to determine this value in the event the unit goes into overspeed, at which time the rotational growth would be greater due to the increased centrifugal effect. This value will help determine the adequacy of radial clearances.

Axial thrust. It is necessary to determine the axial thrust developed on the rotating blades vanes due to steam deflection in the expansion passages. This is determined as part of the thermodynamic design. However, it is normal that the major contributors to axial thrusts (producing loads on both the active and inactive faces of the thrust block) are the piston thrusts developed at the wheels and shaft-end positions.

In reaction units there are no rotor wheels. However, the thrust developed on the blade rows due to steam expansion and acceleration together with the pressure differential that exists across the rotating bade rows will produce thrusts that require special provision to balance.

Diaphragm key loads. Due to steam expansion in the stationary blade passages, there is a tangential thrust developed on the vanes that tries to move the diaphragms in a tangential direction. It is therefore necessary to design a keying system sufficient to prevent such rotation.

Diaphragm support system. When the steam path has been finalized, it is necessary to define a support system for the diaphragms that will maintain vertical and transverse alignment of the stationary to rotating blades under all load and steam conditions. This system is normally arranged to support the lower half diaphragm at the casing horizontal joint. Then the upper half can be located in the upper half casing (often used in the lower pressure stages) or bolted to the lower half.

These calculations are normally undertaken after a final steam path has been defined, stresses and vibratory characteristics checked, and a preliminary layout prepared. At completion of these calculations, or that portion of them required for finalization, the steam path design is either confirmed or minor adjustments made to produce a final design.

The design of low-pressure sections tends to be the application of standard configurations, in which the detailed design of the longer blades is established and cannot be modified without incurring excessive costs. Some minor modification of one or two stages at the inlet to the low-pressure section can be made, but these blades are easily analyzed for both stress and natural frequencies. The design of high and intermediate pressure sections is always individual, with varying energy ranges, number and diameters of stages, extraction locations, sealing arrangements, and blade geometries. The variations and other considerations of these components is considered in detail in following chapters.

SECTION ARRANGEMENT

One important consideration viewed at the time of the initial scoping of the unit is the arrangement of the high-pressure and reheat sections. In these sections pressure and temperature can be high, and leakage quantities represent a considerable loss to the performance of the unit and reduce both efficiency and output.

Therefore, the thermal design engineer will make a decision, normally in concert with the mechanical designer regarding the basic arrangement. Factors that have an impact on the arrangement decision are the flow quantities and the steam conditions. Figure 2–3a illustrates a normal design where the high-pressure and reheat sections are two distinct parts with each section supported on two axial bearings. Each section also has a valves system comprising a stop and a control valve.

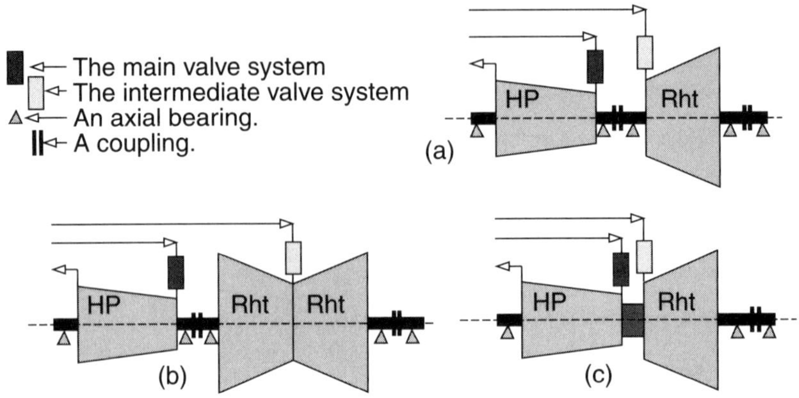

Fig. 2–3 Possible Arrangements of the High and Reheat Sections

In Figure 2–3b, because of increased volumetric flow, the reheat section has been divided into a double-flow arrangement. In this arrangement, the two sections again have separate rotors, are connected through a coupling, and are carried on two axial bearings each. It should also be observed that in arrangement *a* the thrust developed in the two sections will partially cancel so a smaller thrust block is required. However, in *b* the thrust developed in the double-flow reheat section will sensibly cancel leaving the high-pressure section thrust to be carried by the thrust block.

In Figure 2–3c the high-pressure and reheat expansions are carried on a common rotor and will be contained within a single casing. In this design, the inlet to both the high-pressure and reheat sections are located at the mid-span position so the higher temperatures are contained at one location. However, while it is possible to maintain temperature parity at one location, there will be a large pressure differential from high-pressure control stage exhaust to exhaust from the reheat stage stationary blade row. There is, therefore, the potential for considerable leakage at this position, with steam bypassing the high-pressure stages after the control stage.

Consider the unit designed for lower inlet steam conditions and operating within a non-reheat cycle. Such a unit is shown in Figure 2–4. Figure 2–4a shows a design where both the high- and low-pressure sections are carried within a single casing that is supported between two bearings. Often in such an arrangement, the high-pressure section casing is cast steel with a fabricated low-pressure casing attached by bolting. If the rotor span becomes too large, the expansions can be broken into two separate portions, using a four bearing design, as seen in Figure 2–4b. With such a design, it is usual to reverse the flow direction of the high-pressure section so the low-pressure thrust could balance a portion of that produced in the high-pressure turbine.

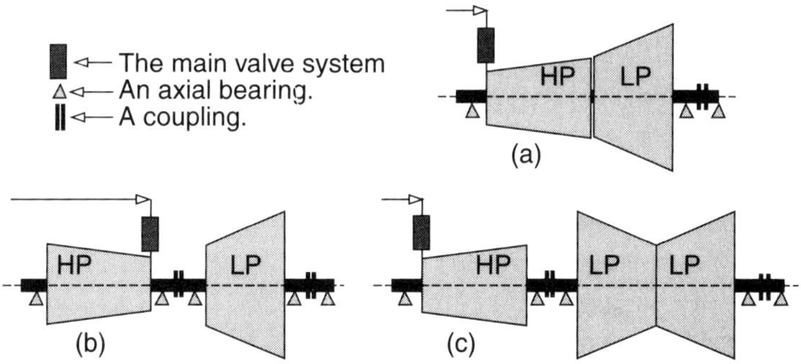

Fig. 2–4 Section Arrangements for High- and Low-Pressure Expansions

If flow quantities become larger or if there is a financial advantage to providing a larger exhaust annulus area, then a two-flow low-pressure design can be used as shown in Figure 2–4c.

There are therefore a large number of configurations that can be used for any application. A major consideration in the selection of a unit by any purchaser is the total economics involved in operating the unit once it is installed and in service. Total costs include the cost of the unit, its efficiency, and the manner in which can be used to meet the demands of customers.

THE MAJOR COMPONENTS OF THE STEAM PATH

It is of interest to review the options available to the designer for the selection of major components. Normally, during any design evaluation and component selection, a number of options will be considered and reviewed before the final decision is made as to which components will be incorporated into the final design.

The following sections will consider the component alternates and their advantages and disadvantages in any particular application. It will also review designs of components that are in operation and have been for many years, but that—with modern practice—would not longer be specified. However, these components continue to provide an acceptable level of service.

THE CASINGS

The casings are the main stationary structural component and are designed to contain the steam path. These casings can operate at high temperatures and pressures, or they can have a vacuum over the inside surface of their outer shell. Such casings are designed to isolate the internal steam environment from the atmosphere.

There are two distinct forms of casing that can be defined by the conditions of the steam they are designed to contain:

The high-pressure, high temperature casing. The high-pressure high-temperature casings are normally produced from cast steel. These casings are designed to contain the steam and, in the event of a rotor rupture, are expected to contain the rotor and any missiles that may be generated. These high-duty casings are normally split along their horizontal half joint with the two halves joined by bolting.

This design is usually produced as a double-shell arrangement with inner and outer portions and the outer-shell completely containing the inner. The gap between the inner and outer shells is filled with partially expanded steam that produces an even steam environment over the entire outer surface of the inner shell and inner surface of the outer shell.

This steam environment reduction is designed to lower the pressure gradient across the shell sections but, more importantly, to divide the total temperature drop from steam inlet temperature to high-pressure section exhaust temperature to more manageable levels. These high duty casings are subjected to high levels of stress as a consequence of the internal pressure of the steam and the thermal gradients that exist across their wall thickness. Thermal gradient stresses can be particularly high as a consequence of fast changes in the thermal gradient within the walls that result from any changes in the operating mode of the unit. Operating mode changes results in a modification of the internal temperature or its gradient through the steam path.

The main components of the casings are the shells—the primary structures In the high duty casings, the shells are produced from cast steel. The casing does consist of a number of other components, principally packing heads designed to carry the shaft-end seal system.

The low-pressure casing. The low-pressure casings contain steam at inlet with a pressure in the region of 170 to 200 pounds per square inch (psi) in a typical fossil application. Discharge pressure is normally produced by a condenser and is therefore sub-atmospheric. In many double-shell designs, the gap between the inner and outer casings is filled with a sub-atmospheric mixture of water and steam that is exhausted to the condenser. This sub-atmospheric pressure produces a large pressure load on the outer shell that the casing must be designed to resist without experiencing any large amount of deflection or deformation that could compromise the alignment of the steam path.

It is normal for the outer shell to be produced as a steel fabrication, but the inner shells can be either fabricated or cast. If the inner shells are cast, then the material used is either steel or, more probably, cast iron. One manufacturer employs a spheroidal graphite cast iron that lends itself to weld repair and upgrading.

In the case of a nuclear unit, because of the higher moisture content of the stream, it is unlikely that cast iron would be adequate to withstand the water-induced damage that can occur. Therefore, in nuclear application it is normal to use cast steel if the inner section is cast.

Like the high-pressure/high-temperature casing, the low-pressure casings consist of several components consisting of the shells and other components.

THE DIAPHRAGMS AND STATIONARY BLADE ROWS

The stationary blade rows can be supported directly from the casing, which has root form grooves produced on the inner surface of the casing of blade carrier. These grooves, exposed from the horizontal joint allow stationary blades to be assembled and locked into the casing or blade carrier. The blade carrier is in effect an inner casing.

The blade elements, since they are manufactured using the same technology as the rotating blades, can be produced to within close tolerances, giving considerable control over the discharge area and angle. This form is a disadvantage if there is any distortion of the casing, causing it to hump or go elliptical. Under these conditions, the stationary blade rows will loose their alignment with the rotating components.

An alternate means of locating the stationary blades is to carry them in a diaphragm. The diaphragm is used in the wheel-and-diaphragm type construction where there is sufficient axial space between the wheels for placement of the diaphragm. The diaphragm consists of an outer ring that is used to locate the diaphragm in the casing, specially prepared grooves, the stationary blade row, and an inner web. This inner web is used to form a radial seal at its inner diameter to prevent the wasteful leakage of steam between the stationary and rotating parts of the unit.

There are various methods of constructing diaphragms and these are considered in a later chapter. Unlike the stationary blades located in the casing, casing deformation will not, therefore, affect steam path alignment, but it is possible for the diaphragm to deform so the potential for deterioration still exists.

The diaphragm and stationary blades are designed to be located in the casings of the turbine and fulfill certain functions. These are

- To carry and locate the stationary blade row so it can receive steam discharging from the previous rotating blade row. This should be done with a minimum of incidence between vane inlet nose and steam angle.
- To form a steam tight seal face where the row locates in the casing or blade carrier.
- To form a steam seal where it is adjacent to the rotor to minimize the leakage losses around the stationary blade row.

Stationary vane design

The stationary blade rows are located in the main stationary portion of the steam path, in the casing directly, in a blade carrier, or as parts of the diaphragm. The diaphragm is designed to form a pressure barrier at the interface between the stationary and rotating portions of the steam path.

These stationary blades consist of vanes specified to admit a specific quantity of steam at known conditions and expand it to a lower energy level converting it to a high velocity jet that is utilized in the rotating blade rows.

These vanes are designed to form an expansion passage between adjacent profiles, and the expansion passage must achieve a certain discharge area and angle.

The nozzle plate

The nozzle plate is contained within the inner portion of the casing. This plate contains apertures for the initial expansion of the steam before it enters the first row of rotating blades. Behind the nozzle plate is a steam chamber that can be arranged to admit steam to the entire circumference of the nozzle plate or to nozzle segments, in which case the inlet valve system is arranged to open the valves sequentially.

There is always some extent of discontinuity at the nozzle plate horizontal joint but in some installations there is a passage from the lower to upper half to ensure an even flow of steam at all load conditions.

THE ROTORS

The term rotor is applied to the major rotating portion of the turbine unit. The rotor consists of a central spindle or shaft and may contain a larger central portion that remains solid (reaction design) or has gashed wheels produced on it (impulse design). The rotor also has blades attached to its outer surface. These are designed to extract the thermal potential energy from the steam. These rotating blades can also employ coverbands and tie wires, if the evaluation by the designer indicates their inclusion will improve either the thermal or mechanical performance of the unit. The rotors may also carry portions of a sealing system that is designed to reduce leakage and prevent steam bypassing the blade rows.

There are various means of constructing a rotor, and details of the form of construction will be considered in a later chapter. However, there are three basic forms:

1. The monobloc, where the central shaft is produced from a single solid forging.
2. The built-up, where individual wheels are shrunk onto a central shaft.
3. The welded, where individual forgings are weld connected to form a continuous one-piece rotor.

There can also be hybrid forms of these three basic types. These will be considered later.

Rotors can be designed to operate in a variety of steam environments and in a number of configurations as directed by the cycle requirements and the design engineer selection of arrangement. Such arrangement is normally reached with the purchaser during the bidding phase of the contract. The following represent the most common configurations.

Rotor configuration

There are a number of possible configurations used for the rotor. There are designs which carry a single steam flow section, those which carry two flows, each flow having similar initial temperatures, those which carry two flows of different temperate and pressure ranges, and those which carry two flows of the same pressure and temperature range. Because of the material requirement imposed on the rotor by steam temperatures there can be different requirements for these different configurations.

Consider the diagrammatic representation of these alternates shown in Figure 2–5. Here we see in Figure 2–5a the single section rotor, where steam at an initial conditions of *P1, T1* is admitted to the steam path, which then expands and leaves the section at conditions *P2, T2*. In Figure 2–5b is a rotor that carries two separate flows. These are at different pressure and temperature ranges but both have the same initial temperature *T1*. The discharge temperatures *T2* and *T3* are not necessarily the same, and are dependent on the pressure drop *P1* to *P2* in the first section and *P3* to *P4* in the second. In this configuration, an intercept stop and control valve is required at entry to the second expansion because steam reheating has occurred and the steam contains a high degree of thermal energy.

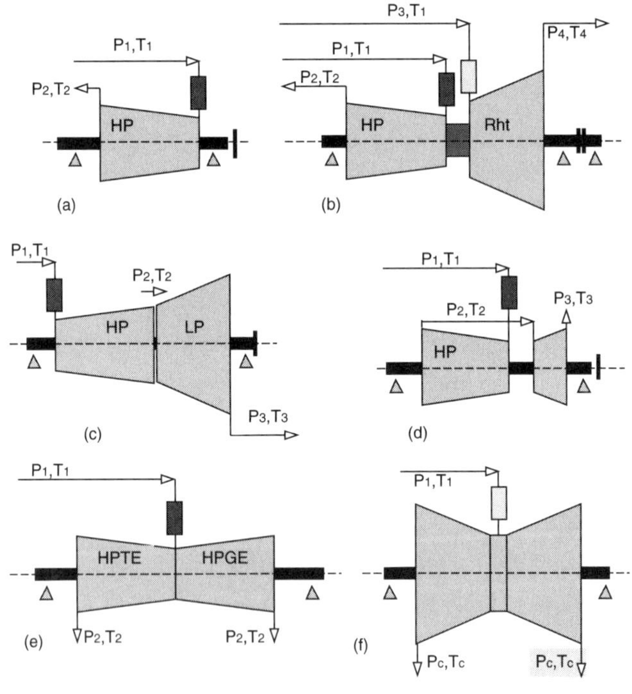

Fig. 2–5 Rotor Configuration Showing Pressure and Temperature Ranges in the Individual Sections

Figure 2–5c shows a design with two expansions—the first from conditions *P1, T1* to discharge at *P2,T2* and the second with an inlet conditions equal to *P2, T2* to discharge at *P3, T3*. From this arrangement, it is clear that the discharging steam from the first expansion is directed into the second where it then expands and discharges to some other point in the cycle. There are several observations that need to be made about this arrangement.

- There is no significant temperature drop from the discharge of the first section to the inlet of the second. Depending on the casing arrangement, there could be a small pressure drop, but the expansion from the first section to the second is basically a throttling operation at constant enthalpy.
- Should the second expansion be a low-pressure section exhausting to a condenser, the volumetric flow of the steam could be such that a monobloc forging could not be used at this location, and a design incorporating monobloc and built-up discs may be necessary.
- Normally the rotor material requirements from high to low temperature are different. There could, although the rotors are supported between the same bearings, be a need to make a bolted connection from the first to the second section.
- If the axial thrust developed in the two sections is high, as might occur in reaction design stages, it could be more convenient for the designer to oppose flow the two sections, as shown in *d*. Alternately, a larger pressure balance piston could be incorporated into the rotor design.
- The type-c rotor form is often used when a three-section low-pressure section is required. This is a practice that is not too common with modern units, but many such designs are still in operation.

- When two flows of the same pressure and temperature range are on a common shaft, there are no concerns regarding the need to change materials. These double-flow rotors are most often in the intermediate- or low-pressure range, although nuclear high-pressure rotors are often double flowed. However, their inlet pressure is at a lower level than that of a fossil design. In e is shown the representation of a double-flow higher-pressure rotor with steam inlet conditions of *P1, T1* and at discharge conditions of *P2, T2*. A double-flow low-pressure design is shown in *f*. This low-pressure rotor has inlet conditions of *P1, T1* and exhaust conditions of *Pc, Tc*. For nuclear application, there would be intercept valves placed close to the inlet, but for fossil application, such a valve system would not be required. Many of these rotors are a built-up configuration with the individual wheels shrunk onto a central shaft.

Technical consideration affecting rotor configuration

The following considerations are the major factors that help determine the size and configuration of the rotors comprising a unit.

Size limitations. Single-piece rotor forgings can only be produced to a size defined by the forge master. However, the ability to produce forgings of greater diameters is increasing with improved manufacturing capability and technique.

The steam condition surrounding the rotor. The steam surrounding the rotor can have a variety of conditions and can contain water or chemical compounds capable of causing various forms of damage.

Rotor materials. Dependent upon steam conditions and intended duty, the material of the rotor will be selected to allow the rotor to operate for the anticipated life of the unit without suffering damage as long as steam conditions are maintained at the design specification and the unit operated within the parameters defined by the turbine builder.

Manufacturing options. The construction details of the rotor are dictated to a certain extent by the experience and manufacturing capability of the turbine supplier. This is recognized by the purchaser at the time of purchase and is factored into his evaluation.

THE ROTATING BLADES

These are the blades that are attached to the outer periphery of the rotor, and they are designed to accept steam entering in a near tangential direction from the stationary blade row, change its direction, and in doing so extract kinetic energy from it. Kinetic energy from the steam is converted to thrust on the blade vane, producing both torque on the rotor and some smaller level of axial thrust. The rotor torque component causes the rotor to rotate and drive some other piece of engineering equipment.

There are two basic philosophies of design—impulse and reaction. In the impulse stage, the entire stage thermal energy is converted to kinetic in the stationary blade row. In the reaction stage, in which the thermal energy conversion occurs in both the stationary and rotating rows. Irrespective of their being impulse or reaction, there are certain categories of blade and certain classes of profile.

The rotating blade consists of a vane, a platform that is adjacent to the vane and acts as a transition piece from the vane to root or fastening, and the root that is designed to transfer all loads developed both on and by the vane and platform to the rotor or wheel.

There are two forms of the vane:

1. **The cylindrical design.** Vanes with a constant profile along the radial height are referred to as a cylindrical design. This type of vane has been used where there is insufficient change

in the steam conditions along the radial inlet to justify the cost of manufacturing a vortex design. However, with the advent of more advanced, cost-effective manufacturing techniques, there is unlikely to be any significant difference in production costs so modern units tend toward the vortex design.

2. **The vortex (twisted) design.** To accommodate changing steam conditions, the vane is made with a constantly changing profile. For longer blades, this represents a twist from root to tip section and allows the shape of the steam passage formed between the vanes to change with radial height to match the changing steam conditions and the degree of reaction that occurs along the radial height in the stage.

The nominal impulse stage is now generally designed with some small degree of pressure drop in the rotating blade row. The ratio of rotating blade pressure drop to stage pressure drop is termed the degree of reaction as seen in chapter 1.

Vanes which approximate the the vortex design. The use of multi spindle and numerically controlled milling machines that are able to follow a master form or computer generated tape has made the production of true vortex blading an economical consideration. Unfortunately, prior to the development of this class of machine tool, the designer used a number of machining devices to approximate the twisted form using planing and plunge milling. Such blades, while not meeting the true requirements of the vortex design, did allow a close approximation, and such blades have operated for many years and provided good service and an acceptable level of performance. These vanes are discussed in a later chapter.

Selection of root form

The rotating blade root or attachment has the primary functions of attaching the vane to the rotor and maintaining the vane in its correct spatial location relative to the stationary blade row and other elements within the stage. It is designed so as not to have stresses—direct or—cyclic, induced in it that could compromise its structural integrity.

There are many forms of root. However, these various forms can conveniently be considered in terms of their main direction of entry to the rotor—axial, radial, or tangential. A brief definition of each follows:

Axial entry. This form normally comprises multi-landed load transfer surfaces arranged in pairs, as shown in Figure 2–6. The root enters the rotor or wheel in a principally axial direction or at some small angle to the axial or, in the most complex form, as curvature to the axial direction. The form of the root platform is dependent upon the vane profile at the root section and the tangential pitch.

Fig. 2–6 An Axial Entry Root with Three Pairs of Load Bearing Surfaces

Radial entry. This form of root enters the wheel in a radial direction and affects load transfer by use of one or more axial entry pins, which pass alternately through fingers produced on the blade and the rotor rim. Such a root profile is shown as Figure 2–7. Here the load developed in the blade is transferred from the blade to the wheel through axial entry pins. On this design there are three pins per root.

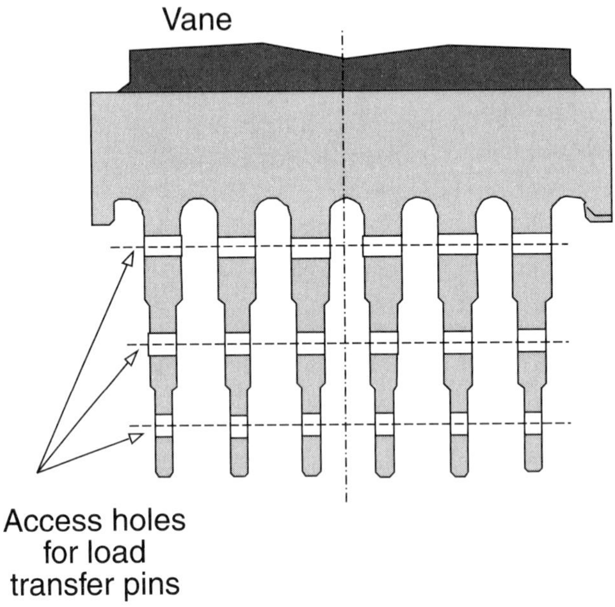

Fig. 2–7 The Root Form for an Axial Entry Attachment

Tangential entry. The tangential entry root, with one exception, requires an access window to gain entry to the wheel rim load transfer surfaces. Two forms of this root are shown as Figure 2–8a and b. Shown in *a* is the inverted *T* form in which the root block is surrounded by the rotor rim material and *b* is the straddle form of this root with the ligaments of the root covering the wheel rim.

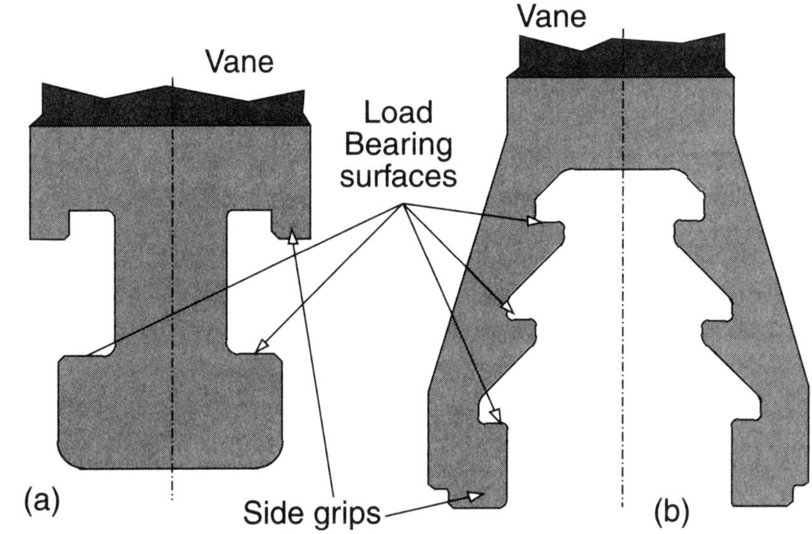

Fig. 2–8 The Tangential Entry Root

THE STAGE HARDWARE

An integral part of the design for a majority of stages is the selection and evaluation of tie wires and coverbands. The vast majority of stages employ coverbands, and some of them are a compound or layered design with more than one layer or band segment. These bands are intended to limit tip leakage but are also used to help dampen vibration. Stages with longer vane blades will also employ tie wires, which are used to tie the blades together and transmit a vibratory movement induced in one to adjacent tied elements. Older designs of long exhaust blades often used three such wires. However, with the more modern blade elements, two wires are normally considered sufficient, particularly where continuously tied structures are used.

Based on previous experience and the natural frequencies and mode patterns of the blade vanes, the number and position of the wires are selected. When preliminary details of the wires have been determined—this can include details of the wire diameter and material—a batching pattern is selected. There are several philosophies used to determine batch size and the manner in which the wires and coverbands are to be arranged or tied. These philosophies require the blades either be arranged in discrete groups with a certain number of elements in each or in a totally random pattern that does not permit tie wire and coverband rows to end at the same tangential position within any row. This random batching tends to make the row more closely resemble a continuous tie in terms of transmitting vibratory stimuli throughout the row. However, this batching does make it considerably more difficult and more expensive to remove blades from the row and to effect partial row repair.

There are two forms of wire the designer will consider. These are the loose and integral types. If the integral type is used, stubs are formed integral with the vane and are joined to adjacent elements by either a ferrule or some metal fusion technique. In the case of the integral wire, the manner of joining the elements depends on the designer and his previous experience with these forms. It is also necessary to consider the potential need to remove the blade elements at some later time. It is possible with ferrules, because there are no solid joints attaching each vane to the adjacent elements and effectively producing a continuous tie. This can be done without imposing the constraint of no tangential movement under the effects of temperature transients and centrifugal growth.

The integral wire is considerably more expensive to produce than the loose wire and, where possible, the designer will specify the loose type. There are intermediate decisions required when selecting the continuous wire, including decisions on its diameter, whether the wire will be solid or hollow, and what material will be used. Will the wire be braze-joined to the vane or left loose with just sufficient discontinuity in the form of sleeves, crimping, or brazed washers to prevent tangential migration during operation.

If a continuous wire is selected, it must pass through a hole in the vane. This hole represents a discontinuous design of the vane with the potential to increase the nominal stress levels in the vane and the stress concentration at the hole. The increase in mean stress levels is introduced because of the reduction in load bearing area at the location of the wire hole. One method of balancing a portion of this increased stress is to add local thickening material in the region of the hole. It is preferable to add area equivalent to that removed by the production of the hole. This is not always possible, but any increase will lower the mean stress level although it may have little effect on the stress concentrating effects. The selection of proper hole-to-vane surface radii is important and has a considerable effect on minimizing the stress concentration effect. The design process evaluates the alternate forms this localized thickening and blend radii can take.

For coverbands, there are two main forms that can be used. There are those formed integral with the vane and those attached by riveting. If the coverbands are to be attached to the vanes by riveting, the design process must select the form and number of tenons that are to be formed as part of the blade manufacturing process. There are two forms (shapes) of rivet in common use—round or shaped tenon. If the blade's outer section is angled, this makes the formation of these tenons a complex manufacturing process. With numerically controlled (NC) machining, this is a relatively easy cutting procedure, but the assembly and production of the coverband section is more complex. The number of tenons specified by the design is a function of the centrifugal load of the coverband that is to be supported and the stresses induced in the cover itself.

The decisions surrounding the coverbands are in part dependent upon the relative axial movement during operation due to differential expansion between the stationary and rotating portions of the unit. It is common practice on the majority of stages to incorporate into the coverband design some form of sealing strip. There are certain variations in the form of these strips that the designer can elect to use.

- The strips can be integral with the cover or they can be produced to form a seal against a platform formed on the cover. In the first option, the radial seal strip is produced on the coverband material, and in the second, the cover provides a radial platform against which a seal located in the casing can produce the specified clearance.
- The strips can form a seal in the axial or radial direction. The choice is a function of the axial movement of the rotor under normal and transient operating conditions. It is normal to use an axial seal when the differential expansion allows this, but for large differential movement stages, the radial seal must be used.
- When produced on the coverband, the seal strips can be produced integral with the band or caulked into it. The main advantage of the caulked strip is that it can be replaced to restore the clearance in the event of rubs opening the design clearance during operation. The main disadvantage of a caulked strip on any piece of rotating equipment is that the clamping effect can be lost during operation. It can become loose and the strip is lost completely. This strip is normally not of sufficient mass to cause significant damage. However, after failure, the leakage quantities increase significantly. For this reason, caulked radial seal strips are usually attached to the stationary portion of the stage.

There are occasions where the tip section of the blade vane will not provide sufficient material for an adequate tenon to be formed. Under these circumstances, it is normal for the vane to be reinforced or thickened locally over a small radial distance. This length must be sufficient to allow an adequate tenon to be formed to support the riveting process and transfer load adequately to the vane.

When the data for the coverband and tie wires is established, there is sufficient information available to permit a details stress analysis of the vane portion and stage hardware. This analysis should be sufficient to ensure these elements are acceptable for the stage.

THE SEALING SYSTEM

The seals used in a turbine steam path are intended to prevent the wasteful leakage of steam that either bypasses the blade rows or escapes from the steam path at shaft-end positions. There are various forms of seal used, but the most common comprises one or more labyrinth strips arranged to throttle the steam from a high to a low pressure. This design is also required to provide a minimum leakage area through which the steam can expand.

Seal system design

Steam seals are used at three locations in the steam path. Seals are also used at those positions where steam is admitted to the casing. These steam inlet seals are not of the labyrinth form and must be designed to be capable of resisting leakage of steam at higher conditions than experienced within the steam path.

The major locations on the steam path are

- Those positions where there can be either a leakage of steam to the atmosphere or where the atmosphere can be drawn into a vacuum location in the steam path. These locations exist at shaft-end positions where the rotor penetrates the casings.

- Where steam can leak under the diaphragm or stationary blade rows, bypass the stationary vanes, and not have its thermal energy converted to kinetic. This leaking steam reaches the discharge pressure from the stationary blades but no ability to do work on the rotating elements.
- Where steam can leak over the rotating blade tips, bypass the blade rows, and do no work in the process. The leakage quantity and output loss can be determined with reasonable accuracy for each location in terms of the steam path geometry and the steam conditions at inlet to and discharge from the blade row.

What cannot be established for the steam that bypasses the stationary and rotating blade rows is the interference effect that exists when the steam reenters the main steam path flow. This reentry steam has only a relatively small velocity and a higher enthalpy. It will mix with and disrupt the normal flow of the steam from blade row to blade row. The enthalpy of the mixture of expanded and leaked steam will be higher than that of the expanded steam alone, so it reduces the available energy entering the next blade row.

THE DEVELOPMENT OF NEW PRODUCTS OR PROCESS

Situations arise where turbine manufacturers undertake development projects and devote engineering resources to these. Such projects can be undertaken over extended time periods or relatively short ones to resolve a particular problem. In fact there are two situations that require the implementation of a development program. These are

1. The proving or evaluation of a design in the event of innovation. In any such development, a management decision must consider

development costs together with predicted eventual manufacturing costs to determine whether the return on investment is adequate to justify proceeding. If such a development project is undertaken to improve existing products is costing more than the anticipated return on investment, then such a project can be terminated.

2. The development of a new design feature or process required to resolve a condition that is causing or is anticipated to cause the structural reliability and availability of the unit to be compromised at some future time, especially if the efficiency will fail to meet design guarantees.

Such a development can be undertaken as a result of failures or because there are indications that improvements to a particular component are required. In this event, the time required to undertake the work depends on the urgency of the situation.

Development because of innovation

Figure 2–9 shows a typical flow diagram outlining the steps involved in the development of a new product or process to enhance the performance of the turbine—if they are proven to be both technically viable and economically advantageous. In any development project for product innovation, there are a number of constraints—technical, financial and ethical—that should be applied. These steps can include:

For a new product to be applicable, the accurate determination of stresses and frequencies should be established during the early stages of the mechanical analysis sufficient to support continued development.

For many components, until suitable values have been determined by calculation and confirmed by suitable and appropriate testing, other aspects of the analysis cannot continue with confidence. This is particularly true of the manufacturing phase, which is often a late starter and not begun until an engineering specification has been developed. Alternately, the final engineering specification may at times await the manufacturing analysis of existing facilities and process specifications so the engineering specification can be complete. This becomes a cooperative effort between design and manufacturing.

The materials selected, for construction or to complete a process, may often be a compromise between competing requirements of strength and other mechanical and chemical properties. The selection of materials can often have an impact on both costs and delivery if these materials are long lead-time items. The final selection of material is based not only on their mechanical and chemical properties at room temperature but how they will behave and the properties they will exhibit under the operating environment in which they will be installed. If the anticipated operating environment range is large, then it is possible that more than one material will be specified for each condition range.

In such circumstances, it will be necessary to undertake a number of tests in different environments on different materials to ensure total suitability.

The effect that proposed design changes will have on efficiency must be determined. If the modification is made as an efficiency upgrade, than the level of improvement must be established with a high degree of accuracy since this will have considerable impact on owner purchase decisions. If the change is being made to improve structural reliability, then it is entirely possible this will have a detrimental effect on efficiency. The extent of degradation must also be determined with considerable accuracy and the results made available to potential customers.

If necessary, both improvements and degradation should be confirmed by tests both in plant and in early field trials, should this be possible.

An important portion of any development project is the assessment of risk. There are often situations where the new product is sensitive to manufacturing, design, or installation conditions. Those engineers responsible for conducting this design development must take every care to ensure that all possibilities of error are considered and that an adequate safety margin exists in the design.

Each of the steps in the development process shown in Figure 2–9 must be carefully evaluated and their risks assessed before the product is offered for sale. To the greatest extent possible, proving tests should be complete so there is a minimal risk associated with the use of the product or process. Such tests may require the cooperation of an owner who is prepared to assist the manufacturer in the development and analysis of the new product or process. However, this owner has every right to expect full disclosure from the turbine builder of expected results, any concerns the engineers have, and the particular areas that are being investigated.

The total costs involved in the development must be determined. These costs include those first time costs that are not to be included in subsequent contracts but must be recovered within a reasonable time frame. Anticipated learning curve aspects of manufacturing costs must be considered along with the number of contracts required to amortize development costs over an acceptable period.

If any significant improvement in installation—in terms of time or adequacy—is anticipated, then the operators of the turbine must be fully aware of the innovative nature of the product or process and be made aware of any risks, however minimal they may be judged to be. In addition, the turbine manufacturer should try to limit further installations until the initial installation has been proven. This may be particularly difficult for the steam turbine, since access may be limited for a number of years and results may be difficult to judge. However, the initial installation should be monitored to the greatest extent possible for any indications of its performance.

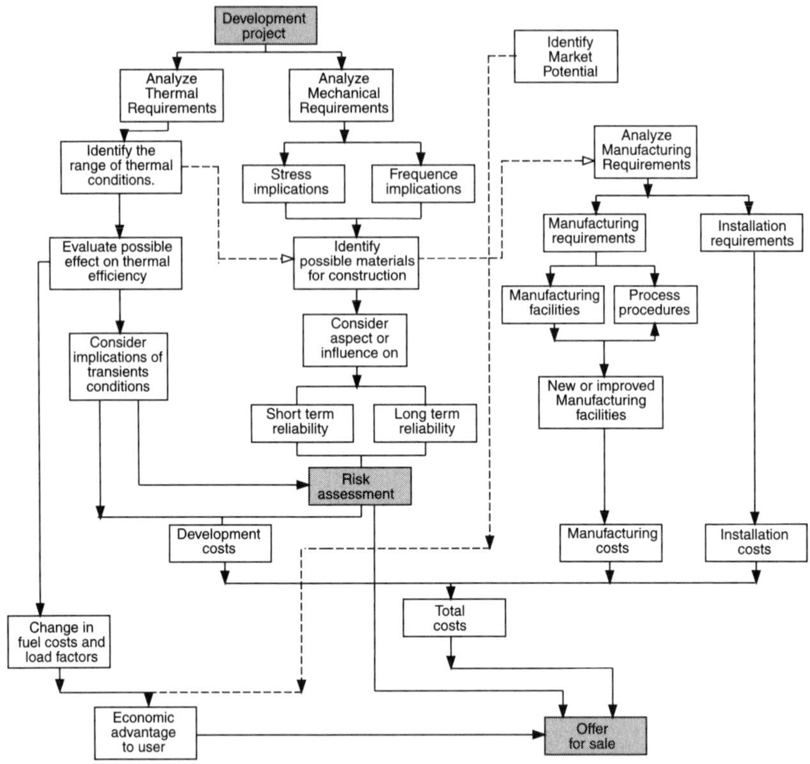

Fig. 2–9 The Evaluation Process for a New Product or Process

When making product efficiency or structural reliability improvements, a first step after identification of the product or process is the identification of market potential. That is, to determine whether the returns are likely to offset the investment required to bring a profit to the company either as a financial return or an acknowledgement of improved product quality for future orders.

In a development for improved unit performance, there are two initial avenues of approach. It will be necessary to develop an envelope of thermal conditions for the region where the product will be applied. This will then allow the mechanical development engineer to evaluate stresses, and, if necessary, frequencies sufficient that suitable materials and processes can be identified. Such identification is required to allow the manufacturing engineer to establish the adequacy of his existing facilities.

The mechanical analysis, in addition to investigating the more obvious characteristics such as stress and frequency, will identify processes that could be required to complete the manufacturing process. The thermal analysis may also investigate the potential for transient conditions to impact upon pressure and temperature excursions and their effect on risk. When such information is available, manufacturing engineers will examine existing facilities and determine whether they are capable of manufacturing the new product or adequately completing the new process to meet the engineering definition of minimum quality requirements.

At completion of the development, the improved product or process will be offered for sale. At this time, risk and thermal advantages have been determined, and the marketing department can evaluate for each potential application the financial improvement this offers to the owner. This improvement in performance is established in terms of current and predicted fuel costs, anticipated load factors, and existing station efficiency (unit heat rate).

Solving an operational problem

If the development is initiated to solve a field operational problem, then the manufacturer does not have the luxury of setting cost and expenditure limits. This is because a condition exists, which may mean that a unit is unavailable and there is a large generating shortfall on a customer system. There are two separate situations that may exist:

The unit has been returned to service derated. In the event the unit could be returned to service, and downtime has been minimized through use of available spanes or the removal of the offending components with some temporary measures made to allow return to service, then time is not as critical. It could be that the operator does not wish, or financially cannot justify, removing the unit from service to incorporate the manufacturer solution, and installation must await the next convenient outage.

The unit is no longer available. If, however, the unit is on a forced outage, then it is incumbent upon the manufacturer to find a solution to the problem as quickly as possible. This could be more difficult than it sounds, and it will begin with a complete design review by the manufacturer to investigate every aspect of the total design, manufacturing, and operational process in an attempt to identify a solution.

There are obviously time limits for developments required to solve problems. In this situation, the cost of the solution is of secondary concern. The manufacturer must make resources available to find an acceptable solution.

If this project is initiated to solve an existing operational problem, then it is unlikely that any technical step can be omitted from the process shown in Figure 2–9. For a reliability issue, the assessment of total costs is irrelevant as there is an existing problem that must be resolved. When a satisfactory solution has been found, there is a need for marketing to sell this solution or improvement to owners with the same problem, either in existence or in the incubation phase.

Such selling is done in anticipation of other users of the same or similar equipment developing the problem and potentially having an outage and possible consequential damage. It can act as a means of recovering all or at least a portion of the development costs.

In undertaking these imposed developments, the manufacturer has little control of costs in any area, and it may be necessary to undertake investigation in several areas or aspects of the problem at the same time, which could increase the cost of development. However, there is little option.

In terms of the overall project there are certain considerations that influence the project. A few of them follow.

- In some design or redesign situations there can be competing conditions between structural reliability and efficiency. That is, the improvement in one aspect should not compromise the adequacy of the other by an amount more than can reasonably be accommodated.
- The development should solve the problem but at the same time consider other regions where similar problems may be developing.
- As an expediency, some portion of the work could be undertaken by outside specialist suppliers of equipment, material, or processes. These should be immediately involved if it is found that a solution can be made if external resources are used.

THE FINAL DESIGN

The final specified design will represent a balanced selection from among a number of component alternates. The ultimate unit, as defined by the mechanical design engineers, will represent choices that have been made to provide a unit that will perform with the maximum reliability, can be maintained, and provide the specified power generated at an efficiency level consistent with the levels defined by the manufacturer in response to the initial inquiry issued by the purchaser, and possibly modified by negotiation during the bidding phase of contract negotiations.

In reaching a final design specification, the turbine manufacturer will make certain compromises. With modern units every area of compromise will be carefully evaluated, sufficient that the final performance of the unit will not introduce an unacceptable level of risk to structural integrity, efficiency, or overall product quality.

During the bidding phase of any contract, the turbine designer will make a preliminary analysis of the requirements of the unit specified, and than make a preliminary choice of section arrangement and the various components the unit will contain. Such a choice is made based on other units of similar design. The purchaser will be generally aware of the design of the various offerings from all bidders, their arrangement, and component designs, and so will not expect to receive a unit with prototype designed elements unless he was aware of these aspects of the bid prior to placing an order.

Stage mechanical analysis

When details of the mechanical design of the various components have been established, the stages and major components should be examined for the stresses induced in them. For complex components, this may often require an analysis be made by finite element methods.

Finite element methods are capable of examining in detail stresses due to centrifugal loading, steam, and vibratory effects. Even thermal effects can be considered, if it is felt these are severe and could under certain situations of operation produce conditions beyond normal expectations. The advantage of these computer methods is that they are capable of more accurately determining the actual stress levels.

The normal hand methods, including those applied through a less sophisticated computer program, determine average stresses. While these may be modified with the application of stress concentration factors to provide an accurate indication of the maximum stresses a component may experience, it is considerably more accurate to determine the actual values from the application of finite element computer methods.

Similarly, for rotating blades, the vane, coverband, and tie wire can and should be evaluated. It is necessary to determine their acceptability from both stress considerations, and, in the case of the tie wire, in terms of ability to dampen blade vibrations. It is not normal for a manufacturer to investigate each stage in detail, since this can be a long and expensive process. However, if there are any questions concerning a new design element, feature, or other non-standard aspect of a stage, then it is far less expensive to make such an analysis than to have the unit forced from service at a later time.

In addition to stress levels, the blade elements require analysis to determine their acceptability with regard to vibration considerations and their ability to operate within the steam environment, subject to impulses generated from the nozzle passing frequency or any other flow abnormality within the stage.

There are various methods available to determine the natural frequency of a blade. The normal method is calculation by hand or computer. The calculated values can then be compared with the anticipated stimuli within the stage. There are many natural frequencies for any blade element because a blade structure has various modes of vibration. It can move tangentially, axially, torsionally, or a combination of ways for both in-phase and out-of-phase relationships. The design must review these calculated frequencies for their fundamental values and the range of harmonics considered significant.

Depending upon the calculated frequencies and the length of the blade, the manufacturer can elect to run frequency tests to establish agreement between the calculated and experimentally determined values. This is, in effect, a calibration method for the calculation system and, because of costs, is normally undertaken only on the longer low-pressure blades, where tuning is essential. However, frequency response tests are also run on first stage elements that are subject to partial admission effects.

There are two methods of determining blade frequencies—using a stationary test rig to determine the standstill frequency or using a rotating rig. This rotating test will determine the natural frequencies of the blade at operating speed where there is the effect of centrifugal stiffening. This stiffening tends to increase the natural frequencies of the elements for most modes of vibration.

When the designer has determined the frequencies of the blade, making allowance for the stage operating temperature that acts to modify the material properties of the blade and its natural frequencies, then the frequency of the blades are plotted on a Campbell diagram and examined for coincidence with the harmonics or per revolution (per rev) stimuli for the stage. If there is no coincidence, then the blade is considered acceptable.

Design review

When the rotating blade elements and the stage hardware are selected and the geometry finalized, the design of the blade is considered complete—subject to a satisfactory analysis of these parts for stress and vibration. No further design analysis is contemplated on these blade elements. However, at that time, the adequacy of the design is reviewed in terms of its acceptability and its ability to relate to and interact with other elements of the steam path. The normal method for making this review is through a design review, at which time the elements are considered for compatibility and their effect on the total unit in terms of speed transients, critical speed differential expansion, rotor deflection, etc.

The final analysis or design review considers all components and their ability to interact with other elements or components both within the stage and the unit. At completion of a satisfactory design review, a manufacturing specification is prepared for each stage blade.

Manufacturing specification

The final act of the designer is to prepare a manufacturing specification. This specification will provide the production department all the information required to allow a blade to be produced that conforms to the design specification. The information is supplied at several levels and by different information channels. In a mature manufacturing facility, many of these exist as standard procedures or methods, but the design should identify each for the blades. This required information includes the following items.

Dimensioned drawings. These drawings will specify the dimensional requirements of the blade vane and root while providing sufficient information to allow the blade to be produced to conform to design requirements. The drawings will detail tolerance requirements and surface finish at all sections of the element. There will also be drawings showing the requirement of the lacing wires and coverbands, including their batching.

Material specification. The material to be used will be specified by the designer. This will normally be a standard for the manufacturer that includes information on chemical constituents such as trace elements, mechanical properties required, tests the supplier is expected to make upon the material, its method of manufacture, and possibly potential suppliers.

Method of material manufacture. The method of material manufacture will possibly be contained in the material specification. If not, this information will be provided separately and will include details of the manufacturing process and the form the material will take upon delivery—forgings, bar stock, or other form.

Special processes. If the blade is to be subjected to any special processes as part of its manufacture, such as coatings or shot peening or if the blades are to be joined into groups by welding, the design specification should define these and provide sufficient information the process can be completed and satisfactory blades produced.

Functional tests. If functional tests, such as vibration testing, are to be undertaken, the design specification should detail these, the number of elements to be tested, and the limits that will be applied.

Non-destructive tests. When the blade is complete, it will be subject to certain non-destructive tests to determine material acceptability and compatibility with design requirements. The design specification will define acceptance limits and corrective actions, if any.

The total design specification will provide sufficient information that the blade can be completely manufactured and tested to ensure its ability to meet design requirements, to be assembled, and perform adequately within the unit for its operating life.

Chapter 3

Turbine Steam Path Performance and the Factors Affecting It

INTRODUCTION

The losses that occur within the steam turbine generator can be categorized in various ways. In terms of this work, they will be considered to comprise those that occur within the steam path proper and others. Those represent losses present in bearings, windage losses at couplings, and other similar losses that are generally termed mechanical losses by the manufacturer. There are also generator losses. Within the steam path, the losses can be further categorized and these will be considered in the later sections of this chapter.

A difficulty often faced by operators and those responsible for the maintenance of steam turbines is establishing when a component has deteriorated to the extent it needs to be replaced for reasons of efficiency. When a component has deteriorated structurally and returning the unit to service without remedial action could force the unit from service, the maintenance decision becomes clear. However, deterioration that affects efficiency, with a loss of output and the requirement of extra funds for additional fuel costs, is far more difficult to substantiate and to quantify. Corrective decisions are often delayed to the overall detriment of the unit and system effectiveness.

Decisions about inefficient operation may also be influenced by considerations of unit availability, particularly when power generation is a secondary consideration and process steam is a more immediate concern. This is especially true if unavailability of a unit could close down a major production process.

In the case of structural reliability and the probability of component failure forcing a unit from service, the decision to undertake corrective action is normally quite clear. However, depending on the form of deterioration present in the components that affects efficiency, the situation is often less obvious, and there can be a considerable financial impact in terms of the cost of replacement parts and delay in return to service. The complexities in assessing the total losses due to inefficient operation are considerable.

Therefore, it is necessary for maintenance staff to be able to make a realistic assessment of the extent of damage or deterioration required to justify the expenditure of maintenance funds when mechanical integrity appears to be acceptable. This is difficult to quantify in terms of impact on efficiency.

The object of this chapter is to provide guidance to the turbine owner, from a perspective of reliability and efficiency, regarding the cost factors involved in a replace, repair, or accept-as-is decision when damage or deterioration is found. Basic rules will be provided and formulas discussed for calculating the cost of a found non-conforming condition within the steam path in terms of the degree of deterioration and its potential impact on future generating costs.

It is generally clear—or can be evaluated—when a mechanical damage condition exists and components cannot reliably be returned to service. After such a determination, an owner may elect to operate at risk. This is a deliberate decision, and in making it, the owner/ operator is consciously accepting responsibility for the consequences, in terms of replacement parts and consequential damage to other components.

A major consideration with inefficient operation is identifying and quantifying the source of losses. Output tests can help establish the total magnitude of such losses, but the determination of their sources and individual magnitudes is often difficult, and a detailed steam path audit is required to quantify and assist plant staff in maintenance decisions. Once the magnitude of losses has been identified and quantified, a financial penalty can be assigned to the losses. It then becomes easier to make replace/repair/accept-as-is decisions based on fuel costs, predicted load factors, and other characteristics of the unit operation. There are also considerations related to the availability of replacement parts and special tools and skills that are necessary to implement the corrective action decisions. These must be fully evaluated by operators to determine the most cost effective approach to correction, return to service, and the need and advisability of taking corrective action.

STEAM PATH PERFORMANCE

Before considering the individual elements that determine performance of the steam path, it is first necessary to consider its definition, and review those elements that comprise and influence it. The factors of efficiency and availability were briefly discussed in the introduction to this chapter. However, in that discussion no effort was given to defining or quantifying those factors or to discussing those factors that affect the steam path's overall contribution to the total performance of the unit.

The steam path can be defined as the annular space bounded by imaginary lines extending axially through the inner and outer circumferential surfaces of the stationary and rotating blade passages. For completion, this should also include the attachment portion of these elements and paths through which the steam/water can expand or leak.

Figure 3–1 shows an expansion of the requirements of performance, efficiency, and availability. It can be seen that the basic considerations of these requirements can be further divided to provide a fuller description. In the case of efficiency, there are both short- and long-term factors, and, in the case of availability, there are important considerations that go into the original selection and design of the components to aid in the effective maintenance of the unit.

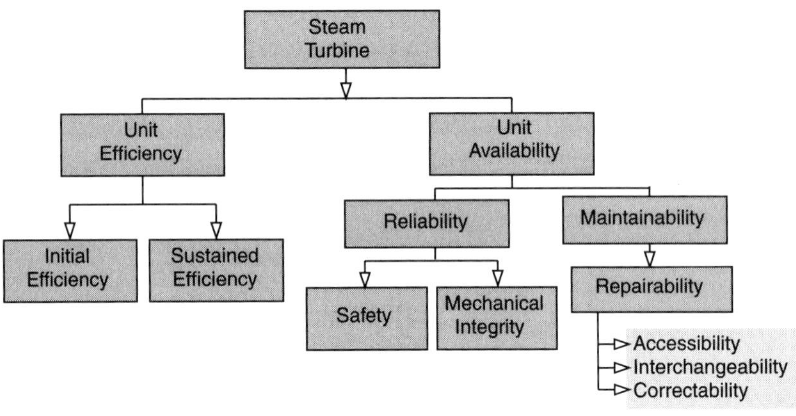

Fig. 3–1 Factors Defining Performance of the Steam Turbine Unit

Efficiency is of importance in terms of both its original or initial value, which represents the design predicted performance, and the value which is most nearly measured in a performance test, and its condition after normal deterioration due to minor mechanical damage, the rubbing and opening of seal clearances and the accumulation of deposits on the blade path has occurred. It is known the initial or mint condition efficiency cannot be maintained, what is then of greater importance is the sustained level, as this more nearly represents the normal operating condition of a well maintained unit, and is also one over which the designer has no control.

Availability relates to the mechanical condition of the steam path from a viewpoint of structural integrity. The unit availability has several sub-elements that must be considered for both short- and long-term effects. The reliability of the individual components, the failure of which can force the unit from service has implications relating to the adequacy of design, and in certain instances can affect the safety of the operating personnel. It is a normal design practice to undertake detailed design stress calculations and select the individual components so they can operate for the predicted life of the unit without mechanical rupture (failure) occurring. This of course assumes the unit has been operated within the design specified parameters, which is an unlikely supposition considering a unit must operate for 200,000 hours within a system with fluctuating demands which must be met by the units. The units as they age, and their efficiency and overall economic viability deteriorates, are often placed on cycling duty, which is considerably more demanding of the individual components and their operating stress levels.

A further responsibility of design is to ensure the individual components are able to be maintained in an acceptable condition, this requires they are accessible, and to the greatest extent possible are either interchangeable or repairable. Interchangeability implies these components can be removed from the unit and replaced, and that in multi unit stations components from one unit are interchangeable with components from other identical units.

A further consideration of performance is the possibility that a unit may under certain damage conditions be derated. That is it may be able to be operated, but with restriction placed on the steam conditions (initial, reheat and exhaust), the quantity of flow entering the unit, and possibly severe restrictions placed on the manner and rate at which it can be loaded and unloaded. Such restrictions can be considered as secondary aspects of performance, with the potential of influencing both efficiency and availability.

STEAM PATH AVAILABILITY

Unit availability is a measure of the unit's capability to generate power when it is required for use by the system or for sale to other users. Availability is comprised of two sub-elements. Reliability is a measure of the unit's ability to continue to operate at the rated load, and maintainability is a measure of the ease with which corrective action can be carried out once the unit has become unavailable.

As a turbine generator unit is built to an engineering specification, it is inevitable that errors will occur in achieving full compliance with the engineering defined optimum requirements. These nonconforming conditions can occur as a result of manufacturing, assembly, or installation problems. In each case, it is the responsibility of the design engineering function to evaluate these conditions and establish the extent to which they can affect the total performance of the unit in respect to efficiency and availability. It can be anticipated that as a unit is put into service it is in a condition that meets with design approval. While it may contain a number of nonconforming conditions, these have been determined, as they occur, to not compromise performance, or, if their occurrence was significant, then the condition was corrected. In operation, similar nonconforming conditions can occur as a consequence of operation and/or maintenance.

Such situations must be evaluated by plant or supplier engineering personnel and a determination must be made of the most expeditious corrective action. When such errors exist, they represent a deviation from the installed (as designed) condition and they must be assessed against the elements or sub-elements of efficiency and availability. It is normally difficult for the operating engineer to know what deviations have occurred, because he is generally not aware of the design requirements. However, some determination should be made of the overall effect of any condition within a unit including its effect on the unit's performance in the short- and long-term.

Reliability levels are a function of the mechanical integrity of the components and can ultimately affect safety. Reliability is a measure of the design adequacy and is influenced by both design competence and the unit's operating history. It is not possible to consider the design responsible for unavailable situations if the unit has not been operated within a reasonable tolerance level of design specifications.

The maintainability of a unit is influenced by certain characteristics of design and manufacture. These include:

- The ability to determine that a failure or nonconforming condition exists. There are certain types of damage or deterioration that can be present, but will not be evident from a simple visual examination or from the more common non-destructive examination techniques. However, methods are being developed to allow damage determination and extent of damage to be defined with a considerable degree of accuracy.
- The accessibility to the failed component and the need to dismantle it from the unit for repair or replacement. Aspects of maintenance are now a special consideration in many design offices, and components are reviewed for interchangeability and ability to be replaced or repaired within short periods of time.
- The interchangeability of components is also a major consideration affecting maintainability. This feature is of particular importance in a multi-unit station where replacement parts

might be carried in inventory for the entire station. If components are not interchangeable within identical units, then this represents a considerable additional cost to the owner, who must then carry a much larger spare-parts inventory. The inventory components of the unit must be dimensionally consistent with design requirements to ensure interchangeability exists.

Long-term considerations of availability require that if field repairs are made, they should be considered in terms of their consequential effect on future repairability, maintainability, and even interchangeability.

Although most parts of the steam turbine do sustain mechanical damage and deterioration, some areas or components have a greater tendency to suffer degradation than others. The more susceptible areas of any unit are a function of many complex factors such as individual stress levels, stress concentration, the mode of operation, and the operating environment. Another critical factor is the operating transients to which components are subjected. This diversity of factors that can influence the potential for damage precludes any generalization of causes being possible or of value. Although no generalization can be made, one area of the unit with a considerable potential to deteriorate and therefore affect unit performance and one of continual concern to the operating engineer, is the steam path and its components.

Many components of the steam path are subjected to high stresses—both direct and alternating. These components may also have regions of high stress concentration because of their complex forms. Also, many operate at high temperature in an environment which may contain chemically aggressive constituents that, under certain regimes of operation, can concentrate to the extent they can initiate failure. In addition to the factors listed previously, it is possible for the stationary and rotating elements to interact with each other in unpredictable ways. All of these factors, when combined with load and steam temperature transients that commonly occur during operation and over which the operator has little or no control, make the steam path highly sensitive and a prime candidate for deterioration.

If a review of any population of steam turbines were to be made, there is a high probability that failures or problems within the steam path would be seen as a significant cause of deteriorating performance for many of them. In particular, the rotating blades are a major cause within the steam path itself, as seen in a recent EPRI study by R. Dewey and N.F. Rieger.

The owners/operators of the steam turbine must develop strategies that can be brought into effect when it has been determined that a condition exists (or is suspected to exist) that warrants the use of an emergency approach to an existing situation. This strategy should include, at a minimum,

- the availability of skilled technicians who can help in the evaluation of the condition and provide guidance as to the most appropriate corrective action. These skills may be provided by employees, equipment suppliers, or independent consultants.
- the provision of specialized capabilities for performing non-destructive examinations. There are only a very few users with the ability to maintain these resources within their own organization. It is possible, however, to maintain a register of companies and individuals able to provide these services.
- the provision of specialized skills, equipment, and materials to undertake emergency repairs. Again, a register of approved suppliers can ease this situation.
- an inventory of those parts required for normal maintenance outages and a means of procuring other more complex replacement parts when these are required.

When it has been determined that a nonconforming condition exists within the steam path, there are normally corrective options available to the user. In these circumstances, it is necessary to evaluate the total condition and make a decision regarding a possible action to utilize properly the technical options and skills that are or can be made available. Decisions relating to the condition must be made and corrective actions selected from those available. The decisions should include estimates of cost and time, but neither of these considerations should be allowed to be the major consideration in selecting the most appropriate action to be followed.

These considerations must also include the potential cost of electing not to take corrective action at the present outage, but to delay repairs to a later time. This option should recognize and include an evaluation of risk and the potential costs of being forced from service before the next planned outage.

The cost of an outage

When a unit is removed from service, there are various costs associated with this action, and these costs must be considered. Such costs are sensitive to many factors and will vary from installation to installation. A brief overview of these factors follows.

Replacement energy cost. When a unit is removed or forced from service, it is normally necessary to replace its generating capacity from other sources. This is, for a utility, generally achieved by starting older, less efficient units. Under these circumstances, there is a differential heat rate between the unit that is removed and the unit(s) started. Alternatively, units may be brought on line with an equivalent or superior heat rate but which consume a more expensive fuel, such as oil rather than coal. There can also be a difference in transmission costs, but these are normally minimal.

When possible, a unit is removed and disconnected from the system when its generating capacity is not required to meet load demands or in the case of a large capacity unit, when a large proportion can be met by utility-owned units. Such periods of time for unit removal are normally in the spring or fall when both heating and cooling demands are at a minimum. The best time for any utility will vary depending on whether it experiences a summer or winter peak.

These various factors of system demand and the cost of running older, less efficient units or units burning more expensive fuel combine to determine the replacement energy cost, and should, if possible, be considered to select a time frame for the maintenance outage when the demand on the unit taken out is anticipated to be a minimum.

Maintenance costs. The maintenance cost has two principal elements—the cost of labor, and the costs of replacement parts.

Labor hours are difficult to estimate with any degree of accuracy and normally are a function of the configuration of the unit and the amount of dismantling and reassembly required. In general, the number of hours required to lift the upper half, remove outer and inner casings, break couplings, lift the rotor and then replace these items after inspection or remedial work, can vary from 1000 to 2000 manhours, depending on the unit, component size, and arrangement. These times can also be increased if any unanticipated faults occur, such as bolts and studs seizing and casings springing. These times do not include any requirements for inspection or remedial work but do include such factors as the removal of lagging and insulation, the disconnecting of valve linkages, and main and intermediate pipe work.

Another element of maintenance costs is the expense of cleaning the parts to make them suitable for non-destructive examination. This cleaning requires blasting the components with some suitable medium to permit the use of a penetrant to detect faults.

Specialized maintenance costs. In addition to the skills required for dismantling and cleaning, which are undertaken either by plant personnel or labor from some local source, or contractor, there is the work that is undertaken by technical specialists.

These specialist skills are either from utility resources or, more commonly, from some outside resource. These specialist activities are in addition to the normal inspection and may be undertaken in response to suspected problems from operational indicators or to meet recommendations from the equipment designer or other technical authority. While these activities can be planned, when budgeting for their completion, they should be considered as additional to normal outage maintenance costs. As units age, these specialist costs are expected to increase.

These specialist costs include labor plus any special purpose equipment that may be used by this labor.

Reserve capacity cost. When a unit is to be removed from service for a planned outage, it is probable this will be done at a period of lower load demand. However, it is normal for other capacity to be made available to meet emergency demands, such as a forced outage on another large unit or the extension of the planned outage to correct some damage that is found unexpectedly and prevents the unit from returning to service of schedule. Similarly, if a unit is forced from service at a period of high demand, then other sources of power must be made available, either by starting less cost effective units or by purchasing from other interconnected systems.

The majority of large utility companies maintain reserve capacity units. The amount of reserve capacity required varies with the forced outage rate of the units on the system and the level of risk the utility is prepared to take in meeting its own system load demands. Another factor influencing the reserve capacity is the annual load pattern. A utility that senses a high peak-to-average ratio during peak load season has different requirements to one that does not. The reserve capacity and the cost of purchasing and maintaining it are therefore important when determining overall power generating costs and the potential cost of a forced outage.

Changes in operating cost. As a consequence of an outage and the remedial actions taken to return the unit as close as possible to a condition consistent with design specified requirements, the efficiency level of an operating unit can change. Most often this change involves an improvement, because the components in the steam path have been replaced or refurbished to improve their overall effectiveness.

In the case of steam path improvement, there will normally be an accompanying improvement in the efficiency of the unit in terms of energy conversion and, therefore, a reduction in the cost of generating power and a saving in fuel costs. These savings can be quantified and represent an operating cost reduction to the owner. What is more difficult to quantify is the period for which such improvements will be maintained, as deposits will again occur and clearances will be increased under the action of rubs that cause wear on the steam sealing system.

There are however, occasional situations when the results of such corrective action will be a reduction in unit efficiency. Such corrective actions are only taken when it is necessary to improve the structural integrity of the unit. An example of this is the removal of a row or number of blade elements that were still in operation when the unit was removed from service but that were in danger of failing and forcing the unit from service.

There are occasions when an outage will worsen the efficiency of power production and increase fuel consumption costs. Such a situation normally occurs only when components have been removed or significantly modified to help ensure they are safe enough to allow the unit to be returned to service. This is usually a temporary measure adopted by the utility until replacement parts can be obtained.

Interval between inspections

A problem faced by the owners of steam turbines is to determine the optimum interval between maintenance inspections. This must be determined for each unit within the system and should reflect operating and repair costs and the consequences of not making inspections or detecting faults in their early stages. Apart from the technical factors, such as the ability to detect a nonconforming condition that can lead to a failure, the following factors influence the time interval between maintenance outages:

- the fuel costs of both the unit being inspected and the unit(s) started to meet load requirements. It may also be influenced by the predicted cost of fuel for the extended outage and the operator's ability to maintain any improvement in generating efficiency.
- the anticipated load factor of the unit during the outages and its output rating. The system load demand will influence these considerations and the point in the year at which the outage is planned.
- the differential heat rates of the unit being inspected and the units being started to meet load requirements. The type and cost of fuel for these units is also important.
- the generic or historic forced outage rate of the unit. If a unit has a poor reliability record, then more frequent outages may be required and easily justified.
- the anticipated maintenance period. If extensive repair requirements have been identified at an earlier outage or suspected from operating patterns, then this may indicate an outage of longer duration.
- known labor costs and replacement parts costs. There is also anticipation of cost escalation for both labor and replacement parts.
- the anticipated improvement in unit heat rate.

- the reserve capacity on the system and the level of risk the owner is prepared to accept in continuing to operate the unit.
- problems discovered in similar units owned and operated by the operator and other utilities. Often the manufacturers of the units will advise operators of potential problems and recommend corrective action and a time scale for its completion.

These factors influence the cost of an outage and the length of time for which a unit should operate between major inspections. Manufacturers do recommend the period between outages because they are the authority with the most significant information on the units, and their recommendations should be given careful consideration in making recommendations for service periods.

STATE-LINE EFFICIENCY

As stated in earlier chapters, in addition to the selection and arrangement of the various components comprising the thermal cycle, a major factor in establishing the heat rate of any installation and/or turbine generator unit is the efficiency with which the thermal potential energy of the steam is converted to kinetic energy and then to mechanical work in the turbine steam path. The efficiency of energy conversion in the turbine steam path is defined as the ratio at which available energy is converted to useful. The most convenient manner of representing this conversion process is to examine the change of steam conditions as the steam expands on the Mollier diagram. The base efficiency of this expansion is referred to as the state-line efficiency, defined as shown in Figure 3–2.

$$\text{State-Line Efficiency} = \frac{\text{Useful Energy}}{\text{Available Energy}} \qquad (3.1)$$

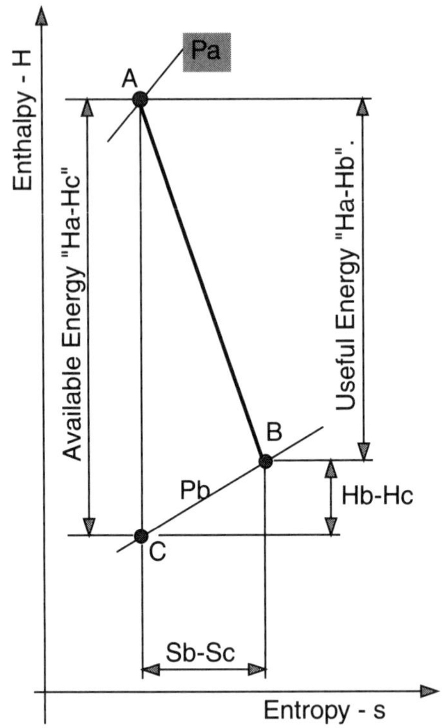

Fig. 3–2 Available Energy A-C and Useful Energy A-B in the Steam Expansion

As the unit operates, any deterioration that occurs within the steam path relates ultimately to the state-line efficiency of that expansion, and it affects the ratio of the available energy of the steam to that converted to useful power within the unit. Therefore, the steam path efficiency can be defined as the ratio of energy per unit weight of working fluid expanding between two pressures that is usefully converted to work in the blade system as compared to the available or isentropic (adiabatic) energy between these same two pressures.

The state-line efficiency for any turbine section is a mean of the individual stage efficiencies, and can be shown on the Mollier diagram as the locus of the steam conditions from stage point to stage point. In its simplest form, the state-line efficiency can be defined from Figure 3–2 as seen in the following equation: (Traupel 1966 and Horlock 1966)

$$\eta_{sl} = \frac{Ha - Hb}{Ha - Hc} \tag{3.2}$$

In this example, the isentropic energy, often referred to available energy, is shown as *Ha-Hc*. However, there are losses in the expansion and a reduction in energy by an amount *Hb-Hc*. The remaining energy is *Ha-Hb*, which is referred to as the useful energy. That is, *Ha-Hb* is the thermal energy that is converted to rotational kinetic energy in the turbine rotor. The lost energy *Hc-Hb* is due to the inefficiencies that occur in the section or expansion. This lost energy increases the entropy from *Sc* to *Sb*.

Within the steam path, the working fluid contains two significant forms of energy. The first is pressure energy released from the steam as it flows through the blade rows from a higher to a lower pressure level. The second form is velocity or kinetic energy. Kinetic energy is produced at the expense of the pressure energy and represents the expansion of the steam. On the Mollier diagram, it is normal to show the state line as represented by the static pressure or thermal potential energy only. When required, the effect of kinetic energy can be superimposed on this potential (expansion line) energy. This superposition is drawn as a vertical distance above the static condition. This is possible because this kinetic energy is available and exists at a constant entropy level. It is also possible that if the velocity is bought to rest the energy level will be reconverted to pressure energy. (In fact, there are losses in the reconversion, but this can be neglected from the present considerations).

Consider the expansion state line Figure 3–3, in which *A* represents the pressure energy at entrance to the turbine section and *B* the pressure energy at exit. At entry to the section, the steam has an entry velocity that, if converted to pressure energy, would have an energy equivalence of *Há-Ha,* and would effectively increase the inlet pressure from *Pa* to *Poa*. The point *Há* is referred to as the total energy condition. Similarly, at discharge from the section, the total energy can be represented by *B′*, with an effective, or equivalent, discharge pressure *Pob.*

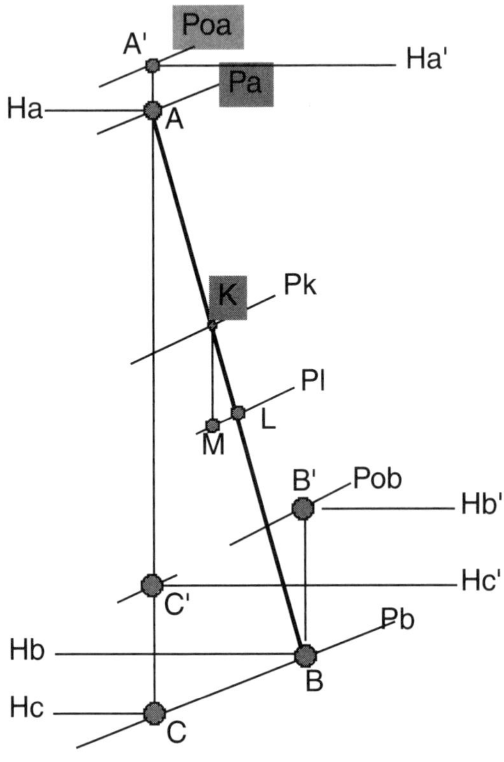

Fig. 3–3 The Expansion Line of a Turbine from Pressure Pa to Pb

If points *K* and *L* on the expansion line at pressures *Pk* and *Pl* represent inlet and discharge conditions for a single stage within the section, then consider the energy conditions around the stage shown as Figure 3–4. Here the inlet kinetic energy is shown as *Hk1-Hk* and is the carryover energy from the previous rotating blade row, and the kinetic energy at discharge from the stage is *Hl1-Hl,* which is the velocity of the steam discharging from the rotating blade row. Using this stage diagram, several definitions of efficiency can be provided.

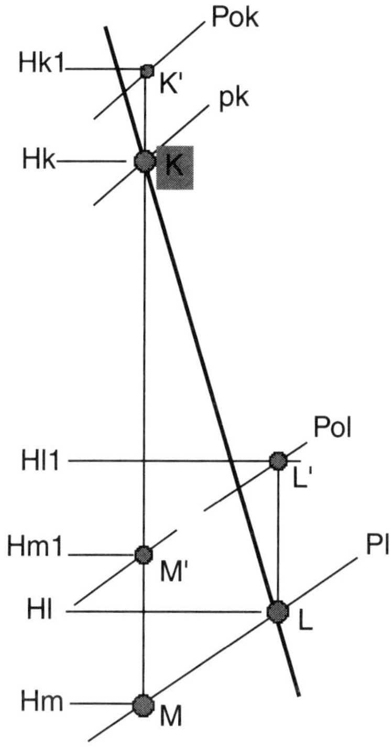

Fig. 3–4 A Single Stage with Steam Pressure Expanding from Pk to Pl

Total-to-total efficiency. The total energy into and out of the stage can be used to define efficiency. This definition takes account of the kinetic energy of the steam at inlet to and discharge from the stage. This is termed the total-to-total efficiency and is equal to:

$$\eta_{stt} = \frac{Hk1 - Hl1}{Hk1 - Hm1} \tag{3.3}$$

Total-to-static energies. A further definition is for the total-to-static energies. This definition accounts for the kinetic energy at the entrance to the stage but discounts kinetic energy at discharge and reflects the stage ability to utilize the total energy entering the stationary blade row. This is equal to:

$$\eta_{stt} = \frac{Hk1 - Hl1}{Hk1 - Hm} \tag{3.4}$$

The first definition, in equation 3.3, is more applicable to a stage in which the leaving velocity is recovered or partially recovered in a following stage. For the total section expansion, if the kinetic energy at exhaust is recovered, then the total-to-total efficiency is as defined in Figure 3–2, and is equal to:

$$\eta_{ctt} = \frac{Ha' - Hb'}{Ha' - Hc'} \tag{3.5}$$

Similarly, the total-to-static efficiency for the section can be defined as:

$$\eta_{ctt} = \frac{Ha' - Hb'}{Ha' - Hc} \tag{3.6}$$

For an exhaust stage in any section, diffusers and flow deflectors are normally used in the discharge region to recover a portion of the kinetic energy. The effect of the diffusers is shown in Figure 3–5, in which the effective exhaust pressure has been reduced by an amount *ΔP* from *Pb* to *Pd.* In this case, the efficiency can be defined as:

$$\eta = \frac{Ha' - Hd}{Ha' - Hc} = \frac{Ha' - [Hb' - \Delta H]}{Ha - Hc} \tag{3.7}$$

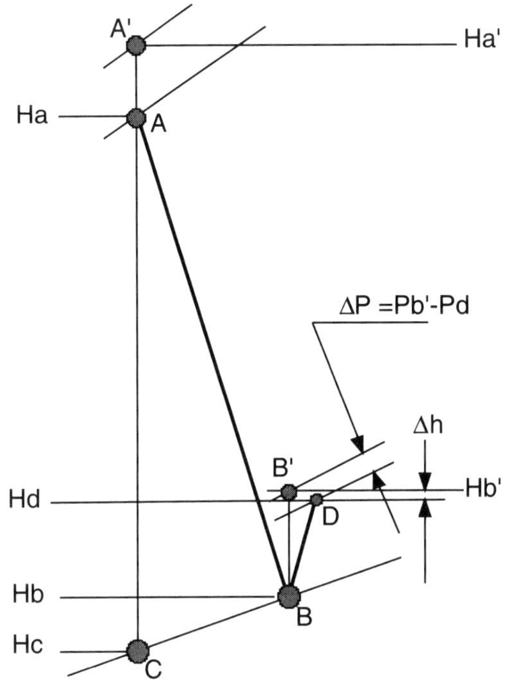

Fig. 3–5 The Effect of Kinetic Energy Recovery in the Exhaust Stage

Efficiency and the design process

The design process selects and sizes components of the steam path to ensure the required output can be developed, the components are reliable, and the steam path is as efficient as possible when operating at design conditions.

In order to begin the design process, the designer assumes a generic state-line efficiency. This is necessary to establish the variation of steam conditions from inlet to exhaust, of the different sections, and the individual stages. Using the estimated expansion line, a stage arrangement is selected, components are identified, and their size and geometry chosen. In addition, stage axial, radial, and tangential dimensions are established.

After the individual elements have been selected and sized but before efficiency confirmation, based on the assumed state line form and flow quantities, the various components are analyzed for structural considerations and reviewed for their potential for failure or deterioration in the environment in which they will operate.

After the mechanical review has confirmed or recommended modification to the selected components and their arrangement, the actual efficiency is determined and compared to the assumed value. If no excessive differences exist, then the selected components are confirmed as suitable for the unit.

The generic efficiency assumed by the designer is predicated on experience and test results from previous units for which both laboratory and field-test data is available. Therefore, the initial state-line assumption is not one made without substantiating data but one that represents known characteristics of the various components and the manner in which they are arranged.

The confirming analysis of steam path efficiency must account for the various losses that can and do occur during the expansion of steam from inlet to exhaust. This analysis considers four basic categories of loss in any stage and/or section that will affect state-line efficiency, as follows.

The stage leakage loss. Some working fluid will bypass the stationary and rotating blade elements through the clearances between them. This leakage represents wasteful or non-expansive work and a loss to the stage output.

Stage moisture loss. After expansion into the moisture region, particles of water are formed in and transported by the steam. Due to the continual acceleration and retardation of the working fluid through the blades, there is a continual drag between the steam and water particles. This causes a loss approximately equal to 1% per 1% average stage moisture.

Stage leaving loss. This includes the kinetic energy loss of the flowing steam together with other losses within the stage.

The basic efficiency associated with stage geometry. This basic efficiency is influenced by vane profiles and blade and stage arrangement. It is also a function of the steam velocity and physical properties.

These losses produce a drop in the available (useful) energy and the power output of the unit. A general expression for these losses in any section in terms of the reduction in available energy, which can be converted to kilowatts, is seen in the following equation.

$$\Delta ht = \Delta hl + \Delta hm + \Delta hv + \Delta ha \tag{3.8}$$

where

Δht = total steam path losses
Δhl = leakage losses
Δhm = moisture losses
Δhv = leaving velocity losses
Δha = steam path fluidynamic and geometry losses

These total losses are represented by the lost enthalpy *Hb-Hc* of Figure 3–2.

In addition to these four types of losses, there are also losses that occur due to radiant heat, bearing friction, and other causes. These are outside the scope of this work but are considered by the designer. (Martin 1913 and Spencer, Cotton, and Cannon 1962)

Stage leakage losses *(Δhl)*

Leakage occurs at any location within the unit where there is a clearance between stationary and rotating components, and a pressure differential exists across them. There are also leakage losses at valves where high-pressure steam exists at stems that must be free to move in guides to control and modify the quantity of steam entering the steam path.

There are four major locations where leakage should be considered in the unit. These are

1. at inner and outer blade diameters on the stationary and rotating rows. For impulse-designed units and many low-pressure higher reaction stages, the inner diameters are in fact at the base of the diaphragms.

2. at shaft-end positions where rotors are coupled to each other in the compound arrangement shown in Figure 3–6 and project through casings to be supported by the bearings. On high- and intermediate-pressure sections, the shaft-end seals stop steam from leaking out. On the low-pressure sections, the glands minimize the inward leakage of air to the condenser.

3. in single-flow reaction turbines where the thrust developed on the rotating blade rows must be balanced by a special provision known as the dummy piston.

4. at the main and intermediate steam valves where leakage occurs, and the steam is led to some convenient point in the cycle.

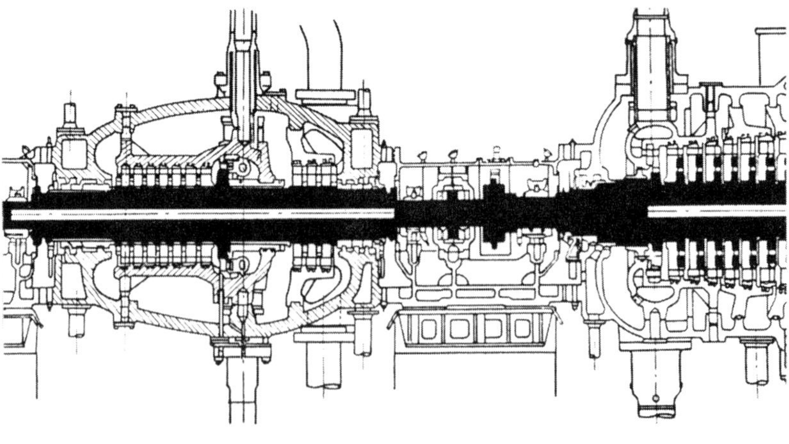

Fig. 3–6 A Typical Shaft-End Arrangement Between the High and Reheat Sections

The leakage points are sealed, to the greatest extent possible, by sealing devices chosen and arranged to limit leakage. In early units, carbon rings located in the stationary components ran making light contact on the rotating elements. Now, a more common system uses a labyrinth seal arrangement that accomplishes sealing by providing a continuous throttling of the steam from a higher to lower pressure through the small clearances between the stationary and rotating parts. These labyrinth seals can be located in the stationary or rotating portions or both.

A general expression for the leakage of steam past labyrinth seals can be found in terms of the equation of Martin. (Baily, Cotton, and Spencer 1967) This equation allows the magnitude of leakage to be determined in terms of design parameters; these include the area through which the steam leaks, the steam conditions, and the number of constrictions in series at the seal location. Consider the single seal constriction shown in Figure 3–7. The steam quantity *m*, in pounds per unit of time, flowing through any area *Ae* can be found from the Martin equation.

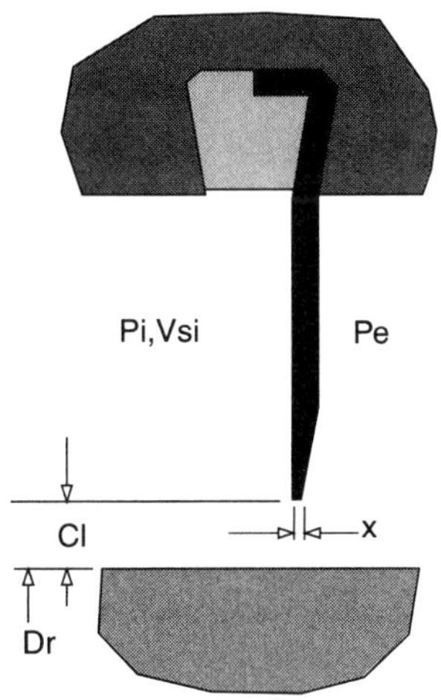

Fig. 3–7 Details of the Single Seal Strip

$$m = k \cdot Ae \cdot \sqrt{\frac{Pi \cdot \left[1 - 1/x^2\right]}{Vsi \cdot (N + \log_e x}} \tag{3.9}$$

Where the leakage area *Ae* is defined as:

$$Ae = 2 \cdot \pi \cdot R \cdot Cl \tag{3.10}$$

where

Ae = the leakage area

R = the mean radius of the seal

Cl = the radial clearance

x = the pressure ratio across the seals Pi/Pe

Pi = the inlet pressure

Pe = the pressure at discharge

Vsi = the specific volume corresponding to Pi

N = the number of series constrictions

k = a constant rationalizing dimensions
in lb/sec = 0.472 and in lb/hr =1700

If more than one series constriction is used, *N>1*, then the Martin equation assumes the pressure ratio *x* is constant across each of the series strips of the total seal arrangement as long as steam discharge velocity past the individual strips does not exceed the critical velocity.

This equation is valid if the steam conditions are in either the superheat or saturated regions for the entire expansion. The leakage flow will reach a maximum when the value of the pressure ratio reaches the critical value. In many portions of the steam turbine, labyrinth seals are used in groups in series with the pressure falling successively through each. At each throttling constriction, a portion of the total thermal energy across the gland is converted to kinetic energy, which is subsequently destroyed in the steam chamber formed between the strips. This kinetic energy is partially reconverted to pressure as its velocity reduces in the chamber. Most of the energy converted at each constriction will be converted to heat or thermal energy.

Each group of labyrinth seals has pressures before and after them as determined by either the cycle parameter or the internal arrangement of the steam path parts. There are two types of flow or pressure distribution to be considered. These are

1. groups with a sufficiently large number of constrictions that the pressure drops across each including the last are less than critical.
2. groups with insufficient constrictions for the energy range so that flow through the last has a pressure ratio x that exceeds the critical value.

Consider the losses associated with these leakages at various locations throughout the unit.

Blade extremities. Within the blade path proper, the inner and outer locations of the individual rows are locations where leakage can occur. Shown in Figure 3–8 are typical stage arrangements. Figure 3–8a shows a stage of the wheel and diaphragm construction and Figure 3–8b the drum rotor construction. In each stage the clearance or leakage areas are determined by the stage geometry. For the wheel and diaphragm design, the inner diameters are considerably smaller than the drum construction, and there is a much smaller leakage area. However, there is a greater pressure drop across this stage and more axial space to assemble series labyrinth seals.

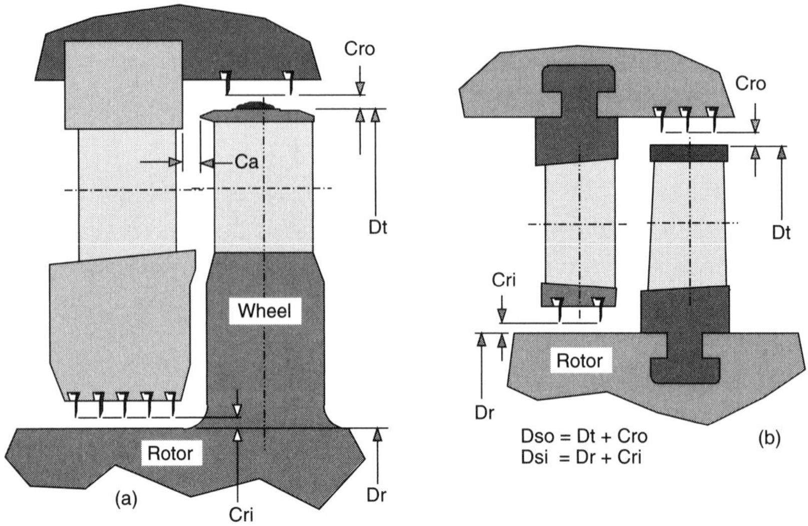

Fig. 3–8 The Sealing Arrangements Made at the Blade Inner and Outer Diameters

The leakage areas for these two stage designs can be calculated as:

$$Aeo = 2.\pi.Dso$$
$$Aei = 2.\pi.Dsi \qquad (3.11)$$

where

Aeo is the outer leakage area

Dso is the outer leakage diameter, = Dt + Cro

Aei is the inner leakage area

Dsi is the inner leakage diameter. = Dr + Cri

As shown in these figures clearances are set at blade extremities. These clearances will have a cold setting as prescribed by design and as measured at final assembly. Under normal circumstances and in the absence of major manufacturing defects, these two values will not be significantly different, and the leakage losses will be close to the design calculated values. Operation has, however, the potential to increase leakage area due to rubs.

Shaft end leakage. Shown as Figure 3–9 is a diagrammatic representation of a shaft-end sealing system with three sealing sections with numbers of seal strips shown as *No, Nm,* and *Ni,* forming two extraction points to pressures *P2,* to some cycle position, and *Pc* to a gland sealing steam condenser. The geometry of the seals is shown the shaft diameters as *Do, Dm,* and *Di.* The clearances are *Clo, Clm,* and *Cli.*

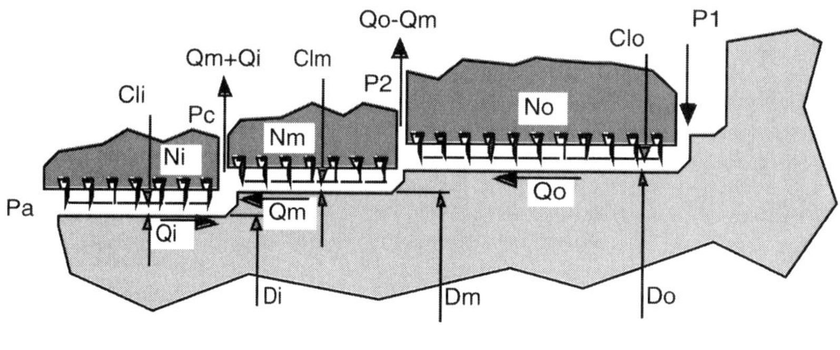

Fig. 3–9 A Typical Sealing Arrangement at a Shaft-End Position

The quantity of steam leaking from the steam path *Qo* is controlled by the dimensions and steam conditions around sealing strips *No,* the quantities *Qm* and *Qi* pass through groups *Nm* and *Ni,* which are controlled by dimensions and steam conditions around the sealing groups. The leak off quantities at pressure *P2* are equal to the difference in the flows through groups *No* and *Nm* or *Qo-Qm.* Similarly, the leakage to pressure *Pc* is equal to the sum of the flow through sealing groups *Nm* and *Ni.* The flow through group *Ni* is *Qa* and will be air. The resulting mix *Qm+Qa* will flow to a sealing steam condenser.

Pressure balance piston. Pressure balance pistons have large pressure differentials developed across them to balance the blade thrust. Figure 3–10 shows a diagrammatic sealing system above such a piston. The sealing system has steam pressures of *Pi* and *Pe* at inlet and discharge. In this design, the seal strips are mounted alternately in the stationary and rotating potions of the seal system. For this reason, the pressure balance piston is normally located close to the thrust block to minimize the differential expansion and allow minimal clearances *Cls* between the strips.

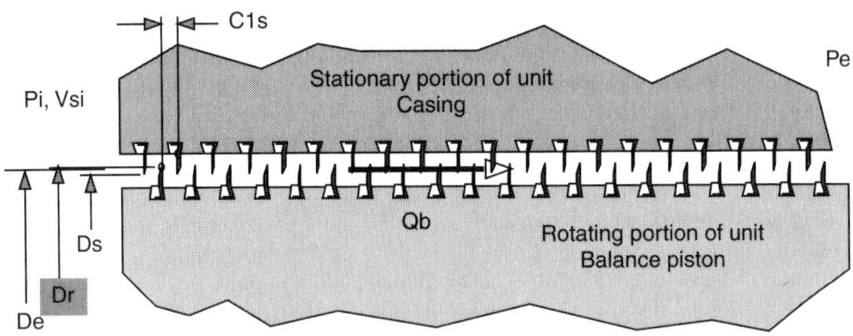

Fig. 3–10 The Balance Piston of a High Reaction Unit

Valve stem leakage. The leakage quantities that escape up the valve stem are collected and passed to some lower pressure location within the cycle. Again the leakage quantity is controlled by the physical geometry of the shaft seal system and the steam properties.

The method of determining the leakage quantities and the power loss associated with this leakage is considered next.

The effect of leakage steam on state-line efficiency

The leakage of steam from one pressure to a lower pressure within the steam path is basically a throttling operation performed at constant enthalpy. However, there is an enthalpy reduction defined by the Fanno Curve. Irrespective of the amount of energy reduction in the leaking steam, the flow entering the main stream after passing through the sealing system is at higher conditions than that which has expanded through the blade row.

The effect of leakage on efficiency is shown in Figure 3–11. Figure 3–11a shows steam that enters a stage at condition *A*, enthalpy *Hi*, and pressure *Pi*. The major portion of this steam then enters and expands through the blade row to pressure *Pm* with a final enthalpy *Hm*. Steam that bypasses the blade row and leaks under the blade inner diameter has a final enthalpy *F* but is at the same pressure *Pm*. Before entering the rotating blade row, these steam quantities mix and produce a mixture condition shown as *C1* with enthalpy *Hc1* at pressure *Pm*, as shown in Figure 3–11b. This condition represents a reduction in the useful energy of *dHs*. The available energy remains unchanged.

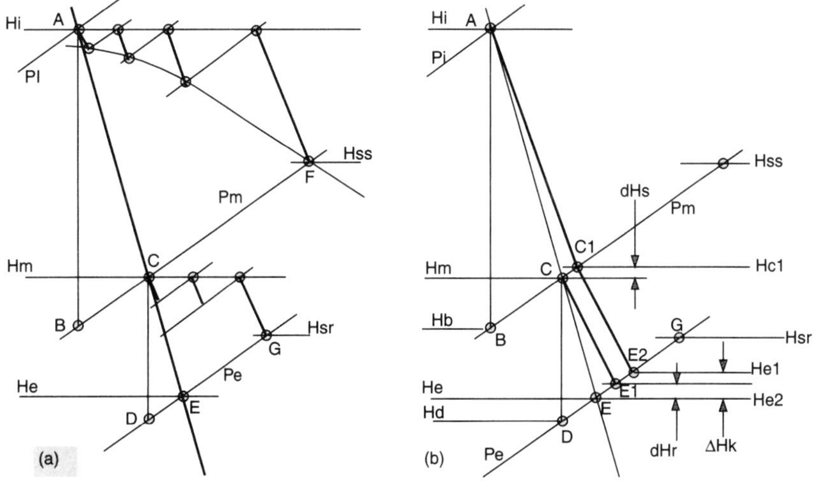

Fig. 3–11 The Effect of Leakage Steam Through the Seals and the Final Condition of the Steam at F in the Stationary Blade Row and Condition G in the Rotating Row

The quantity of steam expanding through the blades is considerably larger than the leakage quantity. Therefore, the final mix condition *C1* will be much closer to *C* than to *F*. However, as the leakage quantity increases due to the wear on seal strips and other damage, the leakage quantity has a more significant effect and the point *C1* will move along the line of constant pressure *Pm* toward *F* making less energy available for conversion to power.

From Figure 3–11a the efficiencies (ignoring the effect of leakage) of the stationary blade ηs and the rotating blade row ηr are:

$$\eta s = \frac{Hi - Hm}{Hi - Hb}, \quad \text{and} \quad \eta r = \frac{Hm - He}{Hm - Hd} \qquad (3.12)$$

When the steam quantities, main flow, and leakage quantity mix, they produce a flow with the conditions represented by *C1* and the state-line efficiency is modified to $\eta s'$ and is equal to:

$$\eta s' = \frac{Hi - Hc1}{Hi - Hb} = \frac{Hi - (Hm + dHs)}{Hi - Hb} \qquad (3.13)$$

Similarly, on the rotating blade row, the condition of the steam entering the row is that of the mixture leaving the stationary blade row. It will have an initial condition *C1* and a final condition *E2* that will have been modified as a consequence of mixing with the steam leaking over the blade tip. This condition *E2* represents that of a mixture of steam at discharge from the rotating blade row. The unavailable energy as a consequence of mixing would be *dHr*. The efficiency of this row $\eta r'$ is then:

$$\eta r' = \frac{Hm - He1}{Hm - Hd} \cong \frac{Hc1 - He2}{Hm - Hd} \qquad (3.14)$$

Had there been no leakage on the stationary row, as occurs on some first stages in the high-pressure section and on the first stage of double-flow units, then there would have been no leakage, no steam mixing, and no degradation of the stationary blade row. Under these conditions, the rotating row initial condition would be at condition *C*, with its final conditions after mixing *E1* at a discharge pressure *Pe*.

For a stage with leakage on both the stationary and rotating rows and enthalpy reductions of *dHs* and *dHr* on the stationary and rotating rows, the stage efficiency ηstg would be:

$$\eta stg = \frac{Hi - He2}{(Hi - Hj)} = \frac{Hi - He - \Delta Hk}{(Hi - Hj)} \qquad (3.15)$$

These expressions do not take account of the reheat effect—an acknowledgement of heat energy produced by frictional heating on the steam path components being returned to the flowing steam and producing a small increase in the stage useful energy.

Stage moisture loss *(Δhm)*

After expansion into the moisture region, water is formed in and then transported by the parent steam through the blade system. Once formed, this water exists in several forms and locations. These are

- As seeds of water formed by nucleation are of small mass, two to three μm, and are transported by the steam. They are continuously accelerated and retarded between the blades as they attempt to achieve parity with the steam velocity.
- Moisture deposited on the internal surfaces of the unit flows with pressure and under the influence of the steam field forces along the internal surfaces from higher to lower pressures. A portion of this moisture is ultimately collected and drained.
- The moisture deposited on the rotating blades is centrifuged from their surfaces to impact with other elements within the steam path.

- A moisture film that is deposited on and flows across the stationary vanes is ultimately torn from the discharge edge of these elements to impact with the inlet edges of the rotating blades.

This water in its various forms does cause damage to the components of the steam path in terms of washing erosion, wire drawing or worming erosion, and moisture impact erosion. In addition to these mechanical aspects, moisture will also degrade efficiency as a consequence of its blunting and kinetic effects.

The small seeds of water formed by nucleation are carried through the blade system by the parent steam. They are accelerated between the stationary and rotating blades to follow the passage shape. These droplets, although small, are of considerably greater density than the particles of steam. Therefore, to achieve steam velocity, they must be both accelerated and retarded by the steam taking a portion of the steam's kinetic energy. This energy then is not available to do work in the stages.

In addition, those moisture particles detach from the stationary blade rows and flow into the rotating blades, hitting their inlet edges and providing a braking effect. The steam loses useful output to overcome this retarding effect.

A considerable amount of research has been undertaken on the effects of moisture content on efficiency, and loss ratios of from 0.80% to 1.20% loss per 1.0% average stage moisture content have been reported. A reasonable value, and that used by many manufacturers, is to assume a loss of 1.0% in efficiency per 1.0% average moisture content in the stage.

The expansion of steam in a unit in the saturated region is shown in Figure 3–12. Figure 3–12a shows four stages, with starting and end points represented on the expansion line by *A . . . E* at enthalpies *Ha . . . He*. In addition, the moisture content at each of the stage points can be taken to be *Ma . . . Me*. Because of the presence of moisture, there will be a reduction in efficiency in each stage, and because the average moisture content is increasing through the expansion, then the degradation will increase becoming more severe as the expansion continues. Figure 3–12b shows in greater detail the first stage, with inlet and discharge pressures of *Pa* and *Pb*. In this stage, the moisture levels are *Ma* at inlet and *Mb* at discharge. Therefore, the mean moisture content is 0.5(Ma + Mb)%.

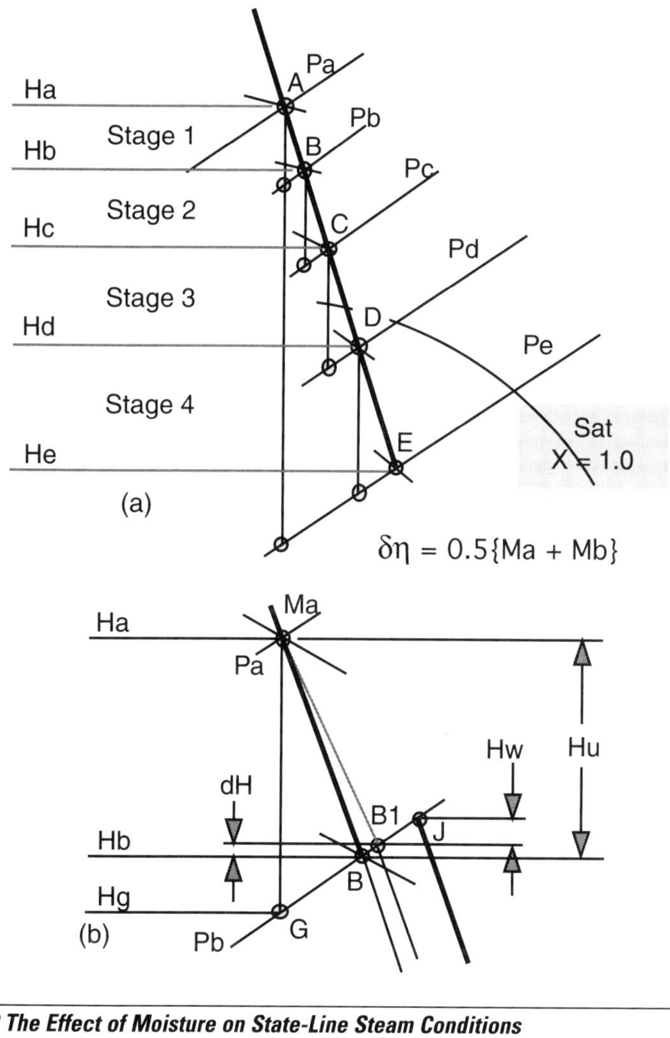

Fig. 3–12 The Effect of Moisture on State-Line Steam Conditions

From Figure 3–12b, the state-line efficiency, discounting the effects of moisture, is:

$$\eta sl = \frac{Ha - Hb}{Ha - Hg} \tag{3.16}$$

If the degradation rate is taken to be 1.0% per 1.0% average moisture content, then the original efficiency is degraded by this amount.

$$\eta sl' = 0.5(Ma + Mb) . \frac{Ha - Hb}{Ha - Hg} = \frac{Ha - Hb'}{Ha - Hg} \tag{3.17}$$

This reduction in efficiency moves the state-line end point from *B* to *B1*, causing a reduction in the useful energy of *dH*. If there is moisture separation at this stage discharge, then there will be a second increase in the enthalpy at discharge from *B1* to *J*, by an amount shown as *Hw*. However, this does not represent a reduction in the output of the stage. It is simply an increase in the entropy and a shift in the state-line position.

The operator can do little to influence this loss beyond helping ensure initial and reheat steam conditions are maintained at the design value at inlet to the unit sections, and where they are provided, water catcher belts and drains can operate correctly to remove the deposited moisture.

Stage leaving loss *(Δhv)*

The total energy available in the steam at any location in the steam path exists in the form of both static (pressure) and kinetic (motion) energy. As steam leaves a blade row, the kinetic energy is carried over to the following row. In the stationary row, kinetic energy is produced from static and directed into the rotating row where the major portion of it is utilized to produce power in the blades. The unused portion is then carried into the following stationary row. Of this carryover velocity, in other than an exhaust stage from any cylinder, a significant portion can be recovered and used in the following row. Such kinetic energy should not be considered lost.

For the exhaust stages in other than the low-pressure units, the exhausting steam will enter the following section with some initial velocity of approach, and could be considered partially recoverable. Because pressure drops of various kinds normally exist between sections, it is considered cleaner to assume all such last stage leaving velocity energy is lost to the system. The energy losses from the high- and intermediate-pressure sections are of relatively small magnitude and are of no great consequence.

For the low-pressure sections, the velocity energy leaving the last or exhaust stage is of large magnitude and cannot be recovered. This energy is called the leaving loss. This leaving loss was briefly discussed earlier in this chapter in defining state-line efficiency.

Steam that leaves the exhaust stages of a low-pressure section represents a loss of energy to the system, and the design engineers always look to minimize this effect. There are several ways to attempt this reduction.

- Providing larger blades and increasing the discharge area to reduce the magnitude of the velocity leaving the blade rows and the velocity (kinetic energy) loss. A major consideration with this solution is that with longer blades their centrifugal force will increase, possibly increasing stress levels and requiring a larger, more expensive low-pressure construction to contain them.
- Providing a greater number of exhaust flows in parallel and reducing the magnitude of the velocity leaving the exhaust stage blades. This makes the axial length of the unit longer and increases the cost of the unit, the auxiliary equipment, and the station within which it is located.

Both of these solutions represent increases in the cost of the equipment and the cost of construction of the entire plant. It is a responsibility of the purchaser to evaluate different options from different bidders and different options proposed by a single supplier to determine the optimum offering in terms of initial cost and the fuel savings represented over the life of the unit.

Other means of reducing this loss include:

- Using a diffuser located at the exhaust from the discharge stage to minimize losses in the flowing steam.
- Constructing the exhaust hood in such a manner that steam is directed into the condenser. This causes a minimum of flow disturbances and reconverts the kinetic energy of the steam leaving the stage back into pressure energy without reducing the vacuum produced in the condenser.

The exhaust loss from any exhaust stage is comprised of three components—the velocity loss, the loss in the exhaust hood, and a turn-up loss. These comprise the total exhaust loss for any stage. These losses are discussed in the following section.

The velocity loss *(ΔHv)*. This is loss resulting from the loss of kinetic energy in the discharging steam. The kinetic energy to be considered is the axial component of the steam velocity leaving the exhaust blades. Figure 3–13 is the vector diagram of the steam discharge velocity from the last stage blade at the mean diameter. A major consideration of design, when a new exhaust blade is developed, is to keep this discharge velocity from the last stage as near axial as possible along the entire radial height—to minimize the deviation of the steam discharge angle variation from the axial δ to as close to zero as possible. This is often difficult, and normally can be achieved for one design load. Normally this is full load or as close to it as possible. Shown as Figure 3–14 is the variation of *Cax,* the axial component of *C2,* along the radial height of an exhaust blade when the unit is operating at full load.

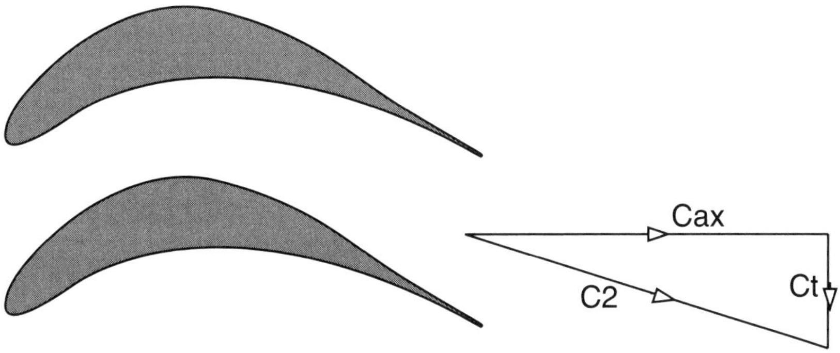

Fig. 3–13 Vector Diagram for the Exhaust Stage

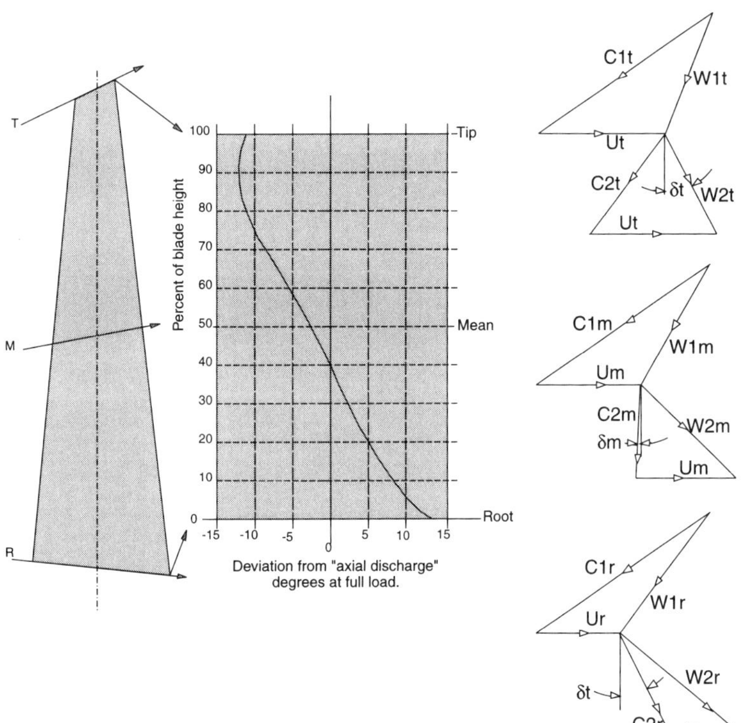

Fig. 3–14 The Variation of the Steam Discharge Angle and Along the Radial Height of the Vane

Figure 3–15a is the loss due to axial velocity for a typical exhaust. As velocity increases, a point is reached at which further increases in velocity produce a smaller increase in the loss. This is the result of the steam velocity approaching sonic. Beyond sonic, the velocity cannot increase. Figure 3–15b shows the hood loss and Figure 3–15c shows the turn-up loss.

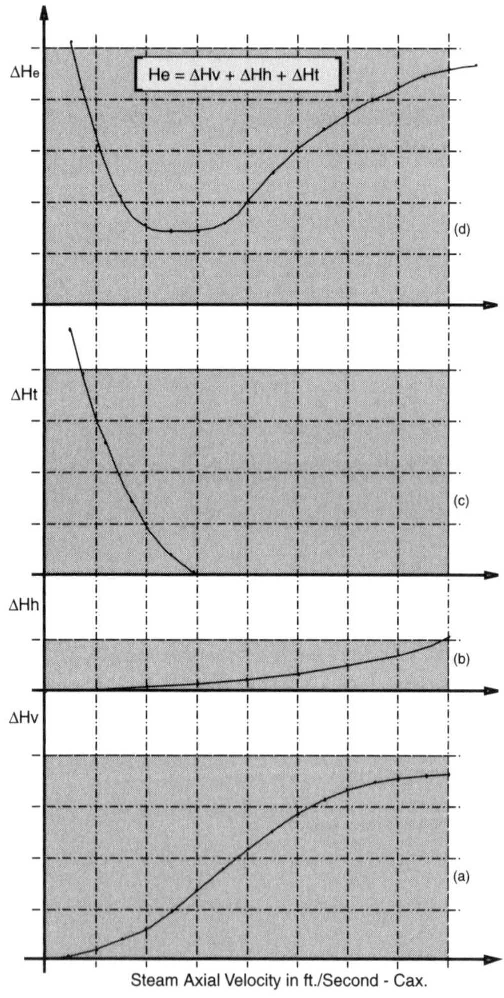

Fig. 3–15 Components Comprising the Total Exhaust Loss from a Last Stage Blade Row

The hood loss *(ΔHh).* The exhaust hood is a vacuum chamber surrounding the turbine exhaust. It must be sized to accept the volumetric flow of steam from the last stage blade, and direct it to the condenser. The condenser is normally mounted below the low-pressure section of the turbine, and the requirements of the hood are to accept this flow and distribute it evenly throughout the exhaust flow-down area. The flow must be deflected to minimize any reduction in velocity that would introduce a pressure rise within the steam and raise the pressure at outlet from the last stage rotating blades.

Shown as Figure 3–15b is the hood loss for the same exhaust stage blade design and hood arrangement as shown in Figure 3–15a.

The turn-up loss *(ΔHt).* As steam discharges from the last rotating blade row, normally it flows through some form of diffuser into the hood. With a well-designed stage, this discharge velocity will be axial or near axial and the exhaust annulus will be sufficiently large to allow the steam to flow into the hood without obstruction. Steam velocity at discharge from the last stage flow is a function of the volumetric quantity, which is a function of the load and/or the exhaust pressure and is particularly sensitive to changes in condenser pressure.

For any specific design of exhaust blade, there is a design point where the efficiency of stage is optimum and discharge is axial. Consider the velocity diagram in Figure 3–16. Here, the optimum design produces a velocity triangle as shown with suffix *a*. The triangle is *RTA*. At this condition, the steam relative velocity leaving the last stage rotating blade is *W1a*, and the blade is rotating with a tangential velocity *U*. Therefore, these velocities provide an absolute axial discharge velocity of *C2a*, which is the same as the axial relative component *C2axa*. If conditions change and the discharge velocity is reduced to *W1b*, then the absolute discharge velocity becomes *C2b* and the actual velocity is not axial but has an axial component *C2axb*. As the steam relative velocity decreases from *W1a* to *W1e*, the absolute velocity changes from *C2a* to *C2e*. As can be seen at these lower relative velocities, the exhaust will eventually choke and the steam will be unable to clear the exhaust, which causes a pressure wave to be established at the blade discharge plane. This effect can have serious consequences on the structural integrity of the blades and will certainly introduce high leaving losses as shown in Figure 3–15c, whereat the lower velocities have a high component of loss. From the characteristic shape of this loss, it is called the turn-up loss.

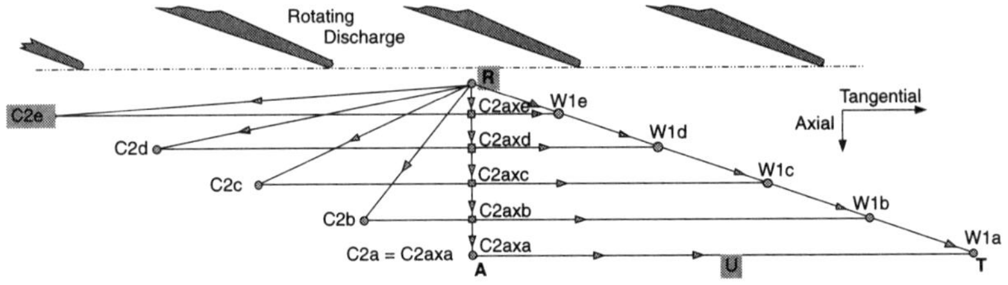

Fig. 3–16 The Velocity Triangles of the Discharging Steam with Reducing Velocities Cax

The total exhaust loss *(ΔHe)* The stage leaving loss or total exhaust loss for any exhaust blade is the sum of these three losses. Therefore:

$$\Delta He = \Delta Hv + \Delta Hh + \Delta Ht \tag{3.18}$$

This leaving loss is obviously a function of the velocity or kinetic energy leaving the last stage blades. However, reducing this velocity does not, after a certain critical value, reduce the leaving loss. As the velocity is reduced, unfavorable values of the ratio of blade velocity to steam velocity *U/Co* occur in the last stage and this poor ratio reducing the stage efficiency. At all loads other than the very lowest, say 25%, the efficiency of the stages preceding the last remain unaltered. This means the pressure ratio across each remains sensibly constant. Therefore, all changes in pressure ratio occur across the last stage.

This worsening of the last stage velocity ratio continues as the exhaust flow is decreased, and reduces the stage output until a point is reached at which the energy loss in the last stage is greater than the stage output. From this load down, the last stage is driven by the upstream stages and any further reduction in flow velocity causes this effect to be transmitted back to the penultimate stage and so on through the last few stages of the low-pressure sections. The cumulative effect of this reduction in velocity is called the turn-up loss.

Figure 3–17 shows the total exhaust loss for typical low-pressure exhaust stages. (Salisbury 1974) If the velocity from the last stage is known, then it is an easy matter to read from this curve the exhaust loss.

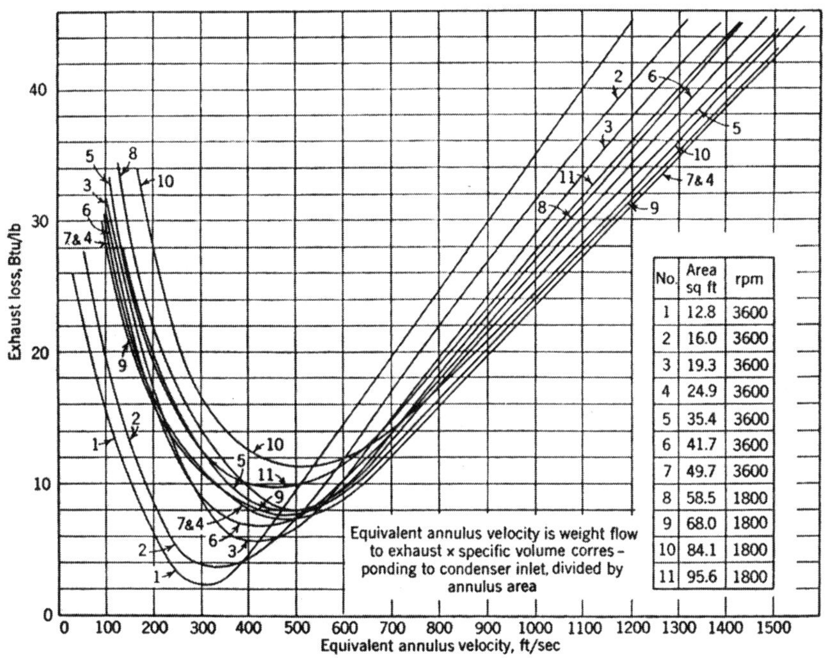

Fig. 3–17 The Total Exhaust Loss Curves for a Family of Exhaust Stages with Various Discharge Areas

It is normally preferable to design the turbine with the last stage blade operating to the right (greater than optimum velocity) of the minimum loss point at full load since the unit will often operate at partial load for extended periods, reducing the leaving velocity. Exhaust area is expensive, and unless the unit will operate continuously at full load, it is often difficult to justify the cost of such additional exhaust annulus.

Basic stage (aerodynamic) efficiency *(Δha)*

Each stage contains two primary groups of components that are fundamental in establishing the basic efficiency of the stage—the stationary and rotating blades. The particular vane profiles used for any stage are selected to achieve primary objectives. These are to be mechanically strong so they can withstand the loads developed on the blade row and to be suitable for accepting the steam by having the correct vane inlet angles along the radial height, and then be able to convert energy from the flowing steam.

Based on the thermodynamic prediction of stationary and rotating vane requirements, the designer selects a profile, normally from among a library of available designs and known characteristics—mechanical and thermodynamic—and specifies their use. The total engineering specification will identify the profiles to be used and the tolerance bands within which they are considered acceptable. The magnitude of these tolerances may reflect either thermodynamic (efficiency) or mechanical (availability) requirements as the limiting or controlling consideration.

These assigned tolerances are selected so the generic efficiency, which was assumed in the initial stages of establishing the state line shape, can be achieved within the scope of the manufacturing processes available. There are certain considerations or characteristics of the profiles and their arrangement that influence the final achieved efficiency. These are considered in greater detail later in this chapter.

STAGE EFFICIENCY CONSIDERATIONS

The efficiency of the turbine stages, and cycle and unit heat rate, are influenced to a considerable degree by the efficiency with which the thermal potential energy of the steam is converted to kinetic energy and then power in the individual stages (blade rows) of the unit. It is known that the efficiency of the individual stages is a function of the efficiency of the rows comprising it, which are in turn influenced by the individual blade vanes and their arrangement within the unit. Arrangement is important in respect to the spacing of each vane relative to the elements within its own row and also from row to row. Typical stage arrangements for both impulse and reaction stages are shown in chapter 1. However, no detail was given in that chapter quantifying or even indicating their effect on efficiency.

In addition to their form and spatial relationships, the efficiency potential of the vane, which is a three dimensional body, is influenced by other considerations relating to geometric characteristics and the use of stage hardware. These also need to be considered.

The following sections will consider stage geometry and its effect on the efficiency of the rows and, where appropriate, the stage. It must also be recognized that the properties of the steam are fundamental to establishing the absolute efficiency of the stage and therefore the unit.

The various methods published for predicting the efficiency of the steam path rely upon the provision of a basic correction factor and then the correction of this resultant efficiency by a number of correction factors relating to the physical and thermodynamic properties of the row.

As an example, the method of Chen states that the total loss is to be correlated with the primary blade geometric parameters and flow conditions. Chen's method is further detailed in "A Loss Model for the Transonic Flow Low Pressure Steam Turbine Blades." It further states that the total profile loss factor ζpo is a function f related by the following expression:

$$\zeta po = f\ \{\alpha 1, P, O, Cf, b, ks, \gamma, Re, \beta 1, \beta 2, M2\} \tag{3.19}$$

where

$\alpha 1$	=	the metal inlet angle of the profile
P	=	the blade pitch
O	=	the throat opening
Cf	=	the skeleton line length of the profile
b	=	the discharge edge thickness
ks	=	the vane surface roughness
γ	=	the ratio of specific heats
Re	=	the row Reynolds Number
$\beta 1$	=	the steam inlet angle
$\beta 2$	=	the steam discharge angle
M2	=	the discharge Mach Number

The method then provides correction factors for the major parameters, and the total loss can be represented as in the following equation.

$$\zeta po = \{\zeta bl \,.\, \chi r \,.\, \chi i \,.\, \chi m \,.\, \chi b\} + \Delta \chi t + \Delta \chi c \tag{3.20}$$

where

ζpo	=	total profile loss factor
ζbl	=	basic profile loss factor with zero trailing edge thickness and zero incidence
χr	=	Reynolds Number correction factor
χi	=	incidence correction factor
χm	=	discharge mach number correction factor
χb	=	discharge edge correction factor
$\Delta \chi t$	=	additional loss due to trailing edge thickness
$\Delta \chi c$	=	additional loss due to trailing edge curvature

Similarly, the method Craig and Cox provides an equation 3.21 for the total profile loss factor. (Craig and Cox 1970–71) This correction provides essentially the same corrections but presents them in a somewhat different form, again defining *_po* as:

$$\zeta po = \{\zeta bl . \chi r . \chi i . \chi b\} + \Delta\chi b + \Delta\chi s + \Delta\chi m \tag{3.12}$$

where

ζpo	=	total vane loss factor
χr	=	Reynolds Number correction factor
χi	=	incidence correction factor
χb	=	discharge edge correction factor
$\Delta\chi b$	=	additional loss due to trailing edge thickness
$\Delta\chi m$	=	discharge mach number correction factor
$\Delta\chi s$	=	additional loss determined as a function of the ratio blade pitch to suction face radius

This method of Craig and Cox considers the total losses within a stage to comprise those associated with the two rows, which are a function of the profile forms and also how these profiles are arranged in cascade. These are termed group one losses. There are also group two losses associated with and more correctly assigned as total stage losses. These groups will be considered next in this text. Also upcoming, curves are provided to show the effect of the various dimensional and steam parameters on efficiency. These curves and correction factors are provided to show the variations that occur and are taken from the various references defined in the text. It must be pointed out that these correction factors were developed by various researchers so they tend to be both profile and unit arrangement specific and should not be mixed. That is, they are provided to shown the magnitude of losses that occur. If it is desired to make a steam path analysis, it is preferable that the curves from one set of references are used and the results of the various researchers should not be mixed.

BLADE PROFILE GEOMETRY GROUP ONE LOSSES

As stated previously, there are a number of physical properties of the profile form, its arrangement, and the thermodynamic properties of the steam that influence the basic state-line efficiency. In this section, this basic efficiency and those factors of the profile and the properties of the steam that affect it will be considered. Various forms of the correction factors that are developed will be provided. The correction factors considered are those listed in equations 3.20 and 3.21.

In considering the basic profile loss factor ζpo and the various correction factors that are applied to it, it must be recognized that many of the dimensional corrections are profile sensitive. They are developed by the researcher for a particular type or family of profiles and are best applied to forms that are within, or physically close to, the type for which they were developed. However, that the profile being considered in any application may be different from that considered by the researcher does not invalidate the information that these various corrections provide. In fact, there is sufficient evidence that these corrections can be used with confidence. An operating engineer is normally concerned with differences from a new to an as existing condition, and therefore error magnitudes in using these correction factors will be relatively small.

Basic profile loss *(ζbl)*

The generic efficiency of any profile is established first by its shape and second—and of greater significance—by the form of the passage between adjacent elements when placed in cascade. Naturally, the passage form, its convergence ratio, and the turning angle are influenced by the profile. Therefore, the profile is designed to achieve the correct form of passage.

To establish a basic profile loss, discounting other geometric variables within the stage, a considerable amount of research has been undertaken by the manufacturers of turbines and also government and educational institutions. In each case, the research undertaken by manufacturers is directed toward a standard family or families of profiles used in their machines.

The research undertaken by universities and government bodies tends to be more embracing of a number of profile families and often considers a larger range of variables. These various results, irrespective of their origin, present some interesting and useful means of efficiency and/or loss prediction.

An important consideration when using available research data to predict the efficiency of any row is that losses of efficiency may be quantified in terms of a number of variables, and the definition given to any particular parameter may vary from one researcher to another. It is essential when using research data that a clear definition of the various parameters be obtained at the outset.

A number of systems or methodologies exist for predicting row and stage efficiency and for taking into account the particular geometry of the stage. However, the primary step is normally to establish the basic profile efficiency in terms of the metal section turning angle and then correct this basic number for:

- profile variations associated with the form and geometric variation of the pressure and suction faces
- losses associated with the shape of the passage formed between adjacent profiles when placed in cascade
- losses associated with the steam properties, flow velocities, and surface conditions

The total efficiency analysis of any stage or turbine section determines the slope of the state line—the ratio of useful to available energy—achieved by determining the extent of energy losses. In total for a section, these represent the lost energy *Hb-Hc* of Figure 3–2.

Predictive methods of establishing these losses are discussed here next.

The method of Traupel. The method of Traupel provides a performance carpet that quantifies the losses that occur for a profile. (Traupel 1966) These losses are defined in terms of the profile inlet and discharge angle. These two angles defining the turning angle of the profile θ, which is determined from:

$$\theta = 180 - (\beta 1o + \beta 2o) \tag{3.22}$$

Definition of the turning angle θ is shown in Figure 3–18 with the profile inlet and discharge angles shown as $\beta 1o$ and $\beta 2o$. These are the angles the profile skeleton line makes to the tangential direction when the vane is inclined to the axial direction at its correct setting angle ξ. This method of Traupel defines losses for various profile inlet and discharge angles and assumes the profiles are pitched tangentially a distance *P* apart in both directions so they form the optimum passage shape and provide a throat *O*. This loss correction does not include geometric corrections additional to this basic value. Such losses could occur if the passage shape was of a converging/diverging type or the vanes were misplaced tangentially or axially to a significant degree. These forms of cascade error are discussed in later chapters. Figure 3–18a shows the basic profile definitions, and Figure 3–18b shows the contraction ratio for the expansion passage from *Oi* to *Od*.

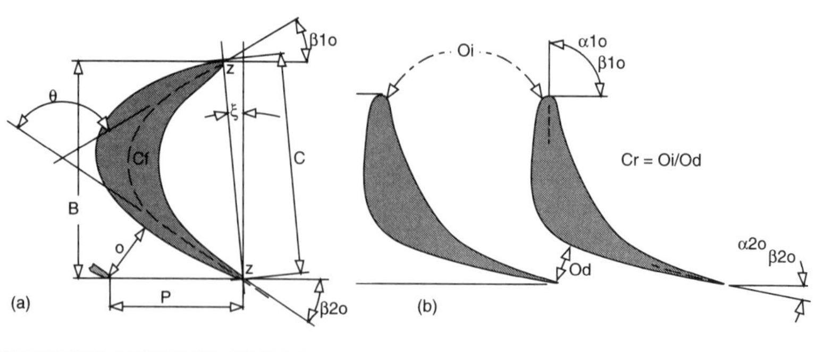

Fig. 3–18 Basic Profile Geometry

Shown as Figure 3–19 is the performance (loss) carpet prepared by Traupel. (Traupel 1966) It can be seen from this figure that predicted losses are defined in terms of the profile inlet and discharge angles. That is, the profile turning angle θ and the geometric data are the information required to define the basic (uncorrected) form of the state line, and the basic efficiency of the expansion ηb can be determined from the simple relationship:

$$\eta b = 1 - \zeta bl \tag{3.23}$$

Where:

ηb = the basic efficiency

and ζbl= the correction factor from Figure 3–19, established in terms of the profile inlet and discharge angles

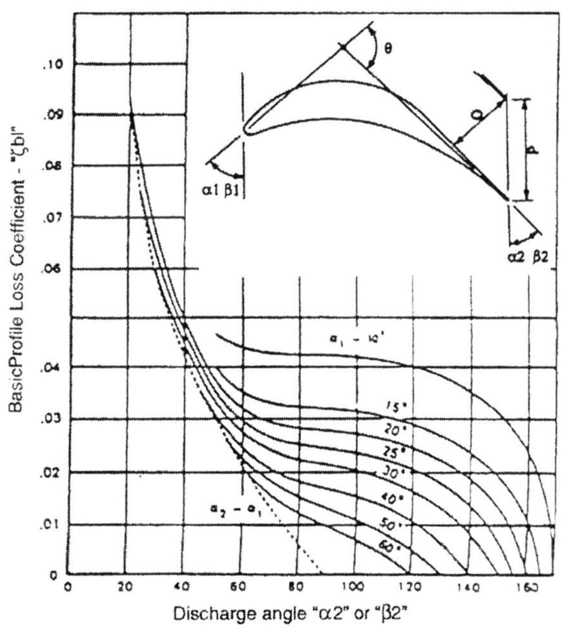

Fig. 3–19 Performance Carpet from Traupel

Using the curves of Figure 3–19 and applying the correction factor *ζbl*, the basic row profile efficiency can be established. The blade angles are shown as *α1* and *α2*, but are also suitable for use with rotating blades with angles *β1* and *β2*. This curve is in a convenient form that permits easy and fast determination of the loss factor. This represents a basic efficiency but one that is uncorrected for profile details and thermodynamic properties of the steam.

The Method of Craig & Cox. From an operational perspective, the method of Craig and Cox is a little more complex and does not lend itself as easily to field determination. (Craig and Cox 1970–71) This is because of the requirement of deriving total geometric definition to allow a field evaluation. However, for design purposes, when design function defines or selects the profile form, such complexity does not exist in establishing this loss factor because the geometry of the profile is known. Shown as Figures 3–20 and 3–21 are the two sets of data required to establish the basic profile loss for this method.

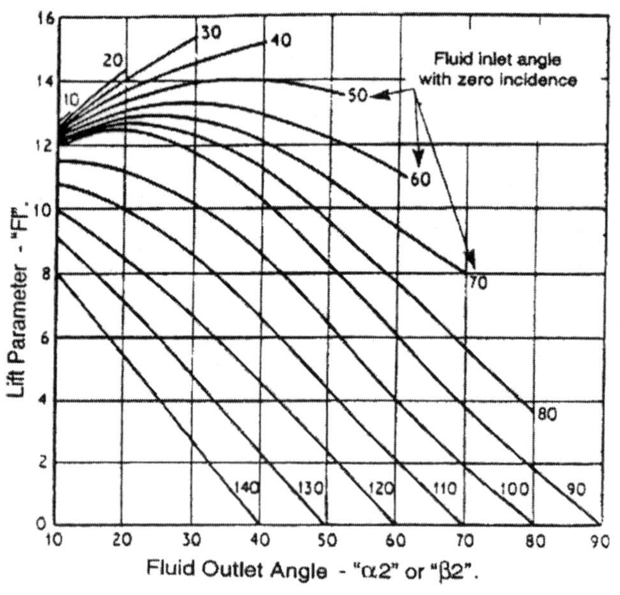

Fig. 3-20 The Lift Parameter F1 for the Vanes as a Function of the Fluid Outlet Angle α2 and β2

Considering the curves of Figures 3–20, this figure establishes a lift parameter *Fl* in terms of the vane skeleton line inlet and discharge angles $\alpha 1o$ and $\alpha 2o$ or $\beta 1o$ and $\beta 2o$. That is again in terms of the profile turning angle θ. These curves assume zero incidence at inlet and the profile discharge angle is the same as the steam discharge angle. Curves are drawn for various angles and it is assumed there is no incidence.

The second correcting factor, shown as Figure 3–21, is the contraction factor. This is the ratio of the throat (opening) width at inlet to the throat (opening) at discharge. A contraction ratio of 1.0 represents a profile that is pitched to provide a constant width passage throughout the expansion, applicable to pure impulse profiles.

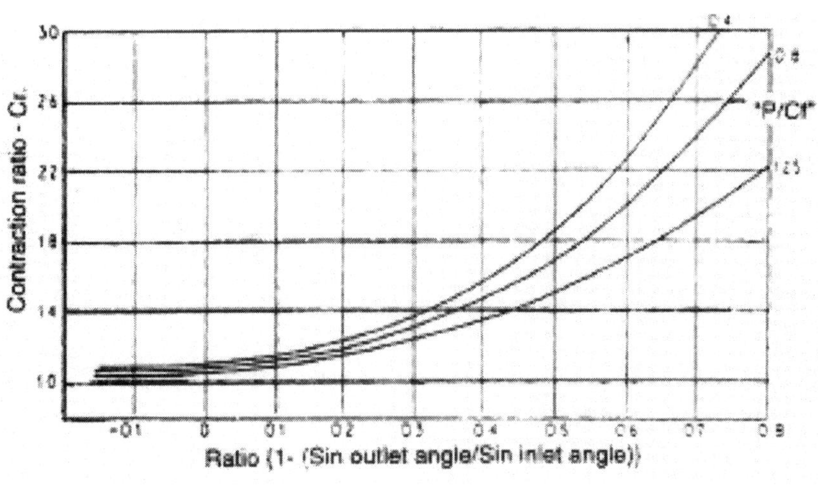

Fig. 3–21 Correction in Terms of the Contraction Ratio

To establish the contraction ratio *Cr*—the ratio at which the expansion passage will diverge—it is necessary to establish the mean inlet throat, shown as the distance *k-k* in Figure 3–18b. However, an acceptable correction can be determined from Figure 3–21, which is a correction in terms of the sine of the metal section inlet and discharge angles. This is determined as:

$$Cr = 1 - \frac{\text{Sin}\ \alpha 2o}{\text{Sin}\ \alpha 1o}\ ,\ \text{or}\ = 1 - \frac{\text{Sin}\ \beta 2o}{\text{Sin}\ \beta 1o} \qquad (3.24)$$

In equation 3.24, $\alpha 1o$ and $\beta 1o$ represent the inlet angles to the skeleton line of stationary and rotating blades, and $\alpha 2o$ and $\beta 2o$ represent the discharge angles the profile skeleton lines make to the tangential direction.

The basic profile loss coefficient ζbl, as defined by Craig and Cox, can then be determined from Figure 3–22 in terms of the parameters of turning angle θ from Figure 3–20 and the contraction ratio Cr determined from Figure 3–21 or making a layout of the total profile.

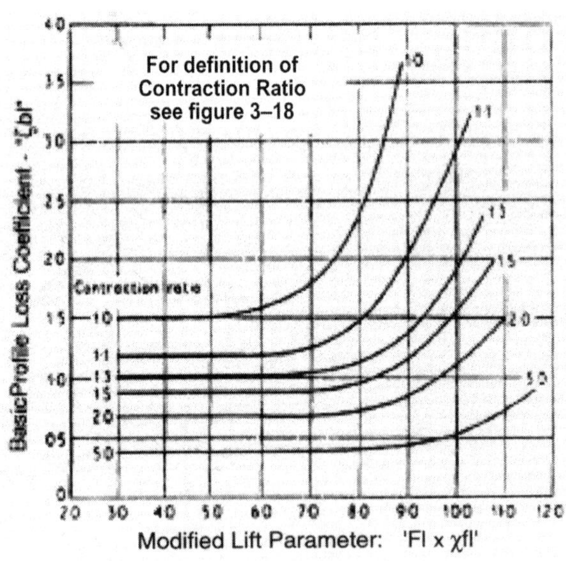

Fig. 3–22 The Basic Profile Loss Coefficient

The lift parameter Fl of Figure 3–20 is modified for geometric ratios of the profile and its pitching. The abscissa of Figure 3–22 is the ratio of the lift parameter *Fl* times the ratio of the pitch to skeleton line length *P/Cf.* In Figure 3-18, the vane skeleton line length *Cf* is defined in as the distance along the curved line *z-z*. The ordinate provides the basic profile loss coefficient ζbl and defines it as described previously.

This method provides correction for variables not considered in the method of Traupel, such as the vane pitching and the passage convergence. However, for field assessment of the profile, it is considerably more complex to determine the loss factor. This method also assumes a discharge tail or trailing edge thickness of zero.

The method of Chen. The method of Chen, developed for transonic profiles, can be applied to all cascades. It allows a correction to be used in the case of steam discharge velocities which are above sonic values. This method establishes the basic profile loss factor ζbl in terms of the profile turning angle θ and is represented by the curve in Figure 3–23. This curve assumes a trailing edge thickness of zero and zero incidence. Therefore, like other correction factors, further refinement of the basic correction must be made for these flow and geometric parameters.

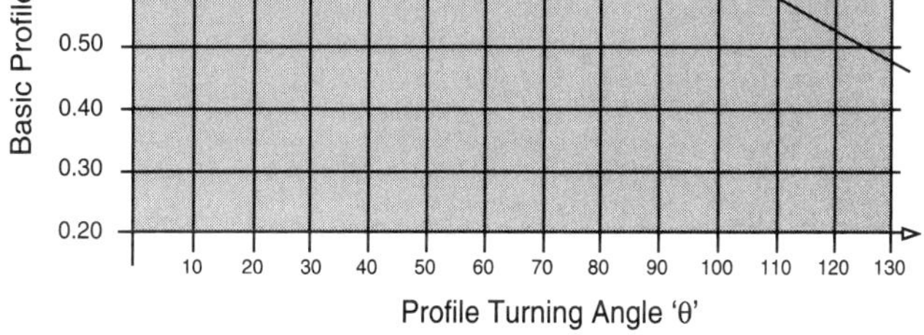

Fig. 3–23 The Basic Loss Factor in Terms of the Profile Turning Angle θ

Methods described by Horlock. Horlock provides information on basic profile loss data determined by Ainley and Mathieson. (Horlock 1966) (Ainley and Mathieson 1957) This data provides performance carpets for profiles in terms of their discharge angle for various ratios of pitch to chord P/C. These carpets are shown as Figure 3–24a for a high reaction profile (nozzle profile) and in Figure 3–24b for a zero reaction profile (impulse blade) in which the inlet and discharge angles are equal.

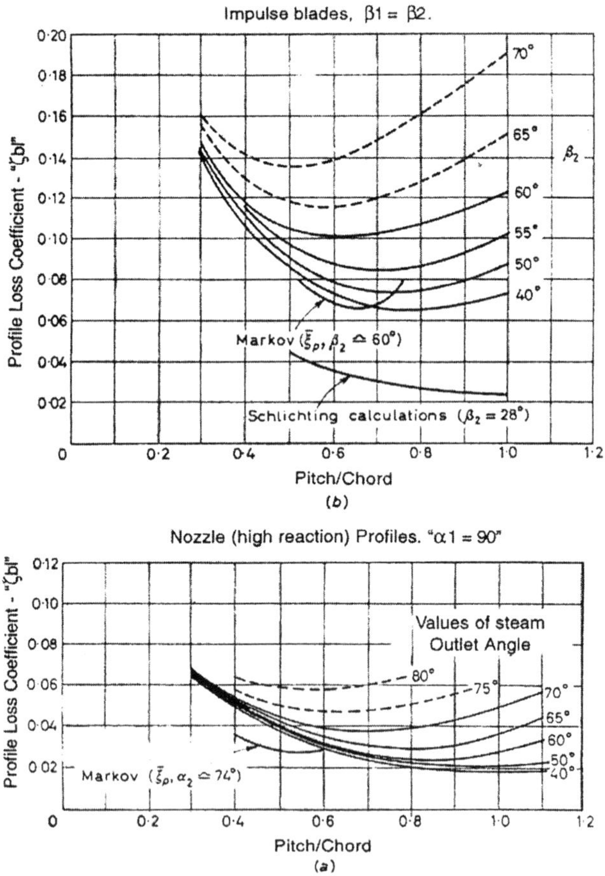

Fig. 3–24 Profile Loss Coefficients for Stationary Type Nozzles for Pure Impulse Type Vanes

In each of these loss carpets, the inlet angle is defined in curve *a* as having an axial inlet, and in *b* as a symmetric profile with $\beta 1$ equal to $\beta 2$. Therefore, these carpets again define the loss coefficient in terms of the profile turning angle θ.

On the carpets provided by Horlock are also shown the results of other researchers, and these values can be found in the same reference.

Each of these methods provides a means of establishing a basic profile loss coefficient ζbl that, if applied in equation 3.23, will allow the basic state-line efficiency ηb to be predicted. However, this value of ηb is not sufficient information to allow the state line to be constructed. There is a need to correct this basic value for other characteristics of the profile and of steam flow parameters together with the flow through the cascade.

Stage Reynolds Number *(χr)*

The Reynolds Number for any row of stationary or rotating blades will indicate whether the flow is laminar or turbulent. Experiment has shown, transition from laminar to turbulent flow takes place at a Reynolds Number dependent upon the magnitude of the disturbance. (Abbot and von Doenhoff 1958)(Milne-Thomson 1958)

For viscid fluids at very low Reynolds Numbers, as in oil, all disturbances are damped by the viscosity of the fluid, and the flow is laminar regardless of the magnitude of the disturbance. For non-viscid fluids, as the Reynolds Number is increased, a condition is reached at which some particular types of disturbance are amplified and eventually cause transition. This value of the Reynolds Number is called the lower critical. Further increases in the Reynolds Number cause amplification for a greater variety of disturbances and increase the rate of amplification. Under these circumstances, the Reynolds Number for transition depends on the magnitude of the disturbance. Transition can be delayed to higher values of the Reynolds Number only by reducing all disturbances such as stream turbulence, unsteadiness, surface roughness, and discontinuities to a minimum.

In general, within the turbine steam path, where the fluid is not at any pressure level viscid, transition can be taken to occur at a Reynolds Number of about 1x10E6. It is unlikely many stages in the high or intermediate pressure range of modern units will operate below the transition point at discharge from the blade row. However, for stationary blades and some rotating blades with a high level of reaction, the steam is accelerated within the steam passage and could therefore be expected to be laminar in the inlet regions of the passage.

Generally, the higher the Reynolds Number at discharge from the turbine blade row, the higher potential efficiency this row has. This is partially due to the fact that at higher velocities the boundary layer becomes thin and there is less disturbance transmitted from the passage walls to the flowing fluid. However, with these thinner boundary layers, the greater potential exists for turbulence and flow separation with a poorer finish on the profile surface. This poorer surface is not only a function of the manufactured surface but also the surface that exists as a consequence of deposits or mechanical damage.

Transition from laminar to turbulent flow can occur in the passage formed between any two profiles. It is important for profiles to be produced so they minimize the possibility of premature transition from one regime to another. This requires ensuring an adequate surface finish and correct blending from radius to radius or a continually changing radius of curvature on the profile surfaces. It is necessary to prevent sudden discontinuities or change of radius of curvature at the profile surface. This is discussed further in chapter 8.

It is obvious that flow separation can be initiated by a number of parameters or characteristics of the profile or passage, including surface finish. Blade manufacturers take considerable care to produce a smooth surface in which the lay direction corresponds to that of the flow. It will also be a surface that can be produced by conventional manufacturing methods, and normal hand polishing. Hand polishing is a process that requires care on the part of the manufacturer and in the worst situation can vary not only from vane to vane, but also over the surface of any particular element.

The surface finish required for any particular blade row to minimize losses can be defined in terms of the stage Reynolds Number, which is a function of the steam conditions and row geometry. A difficulty encountered when comparing these different experimental correction factors is the profile dimension on which the Reynolds Number is based. These differences make direct comparison between the various data difficult, and care must be exercised in establishing the appropriate parameters used when comparing these values.

Early work by Prandtl and Schlichting provided curves for the coefficient of turbulent skin friction for flow across a flat sand roughened plates of length *L*. (Prandtl and Schlichting 1934) Their results are shown as Figure 3–25, and they define the hydraulically smooth surface as being a surface on which the rugosites of the surface imperfections do not project through the boundary layer. From this curve, the value of frictional coefficient *ks* could be determined. This value of *ks* cannot be used directly to correct the basic efficiency coefficient *ζbl*.

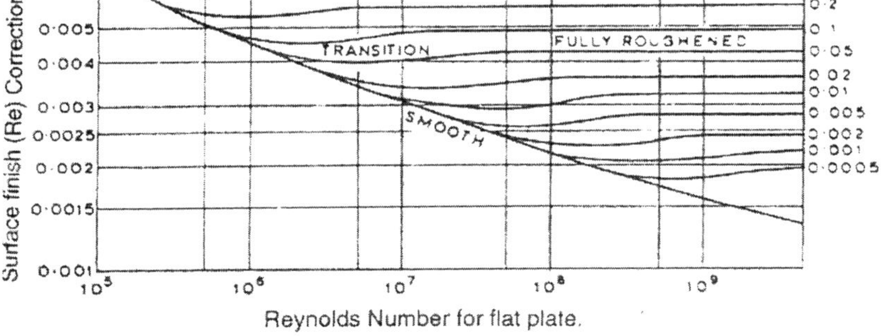

Fig. 3–25 Reynolds Number Correction Factor for Various Surface Finishes

The various correction factors for the ratio of surface finish *ks* to profile geometric parameters as a function of Reynolds Number include:

The method of Traupel. This method provides correction curves that relate a correction factor χr to the ratio of surface finish in micro inches or millimeters to the blade chord *C* (as defined in Figure 3–18a) in inches (or millimeters) as a function of the row Reynolds Number *Re*. These factors are shown in Figure 3–26, and it can be seen from these curves that a lower boundary exists. Below this there is no advantage to improving the surface of the vanes. At these values, the surface is sufficiently smooth that the boundary layer is thicker than the irregularities, and these surfaces will not promote separation or turbulence. The surface is hydraulically smooth.

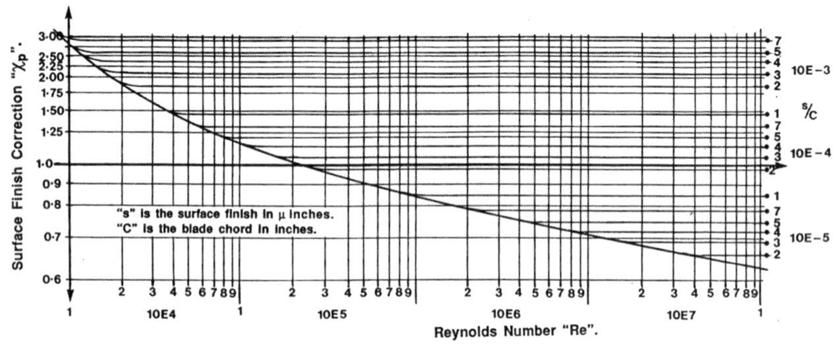

Fig. 3–26 Surface Finish Correction Factor χr as a Function of Reynolds Number for Various Ratios of Surface Finish to Vane Chord Length

The losses associated with the Reynolds Number are those losses that occur in the boundary layer, and those losses associated with separation of the boundary layer from the profile surface. Therefore, it is to be expected that as the surface becomes rougher the magnitude of the losses will increase. This fact is substantiated by the curves of Figure 3–26 where the correction factor χr increases as the ratio *ks/C* increases.

In these curves, the chord used to determine the ratio *ks/C* is the value across the profile horns as seen in Figure 3–18. That is the shortest distance from z-z.

This value of χr is then used to correct the degradation of state-line efficiency and modifies equation 3.23 to:

$$\eta b = 1 - [\zeta bl . (\chi r)] \tag{3.25}$$

These correction factors from Figure 3–26 for surface finish lend themselves relatively well to loss determination in the field because the true intent of field observations is normally to establish changes in loss levels and the level of recovery that can be anticipated. This is, therefore, a problem that devolves to assessing changes in surface finish levels. Unfortunately this surface finish assessment is subjective and calls for mature judgment on the part of the engineer making the overall assessment for the following reasons:

- The surface finish, with the effects of deposits or damage, varies across the entire surface and a mean or effective value must be assessed.
- The surfaces on the pressure and suction faces may be entirely different, and it becomes necessary to assign levels of importance to both.
- The velocity throughout the expansion passage changes positively or negatively, and it is necessary to define levels of influence over the surface depending upon its proximity to the discharge point.
- There can be surface variations along the radial height of the vane, depending on the location of damage or deposits.
- The effects of mechanical damage may be localized or cover a complete surface at one axial or radial location.

The typical variation of surfaces on vanes is shown in Figure 3–27. This shows the uneven nature of an iron/silica deposition on a stationary blade row. Figure 3–28 shows the effect of mechanical damage, which destroys the surface and modifies the discharge area and angle.

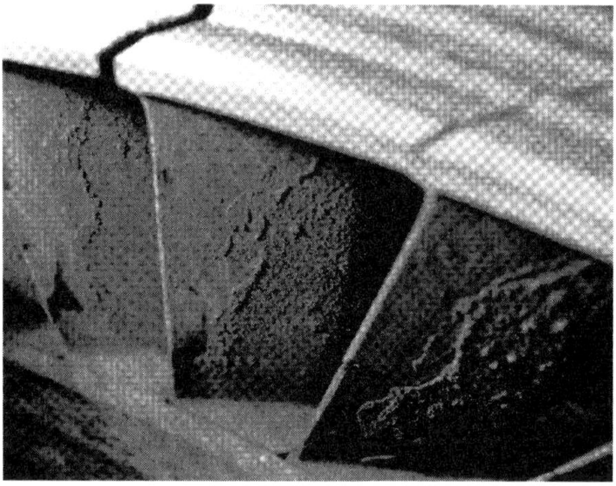

Fig. 3–27 Deposits on a Rotating Blade Row Causing Surface Roughening

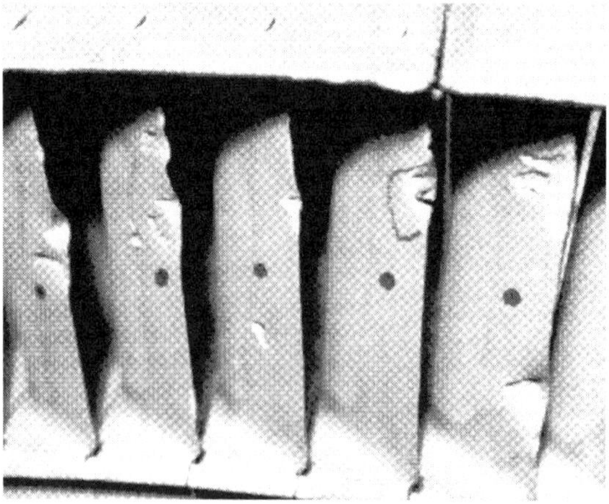

Fig. 3–28 Mechanical Damage to a Rotating Blade Row Sufficient to Cause Flow Disturbances

The surface finish/Reynolds Number correction curves are of considerable value to the design engineer in terms of assessing the surface required of any row and then, if the design required surface is finer than can be achieved by conventional manufacturing techniques, in determining the effect of the manufactured surface on losses. They do present a considerable level of difficulty to the plant engineer in assessing the effective surface when determining the magnitude of field losses and the level of recovery expected from corrective action.

The method of Craig & Cox. The method of Craig and Cox provides a similar form of correction, with factors substantially similar in form to those provided by Traupel. However, in this method, the Reynolds Number for the row is based on the width of throat between the profiles, rather than on any definition of physical characteristic of the profile. For any cascade this provides a suitable means of loss prediction but does introduce a difficulty if there is a significant level of profile change along the radial height, resulting in large changes in the passage discharge throat.

As with the method of Traupel, this does not present any significant difficulty to the designer, but the field engineer has a need to determine mean values. For short radial height vanes, this is not a significant difficulty and values at the mean diameter are adequate. But for the larger vanes, particularly those of a vortex form, significant errors can be introduced.

With the Craig and Cox method, the surface finish correction is determined for the stage Reynolds Number for various ratios of s/C. However, in this case, the value of s is the equivalent sand grain size, and the value of chord is the skeleton line length Cf. This correction factor is shown in Figure 3–29. The surface finish correction factor is Xr.

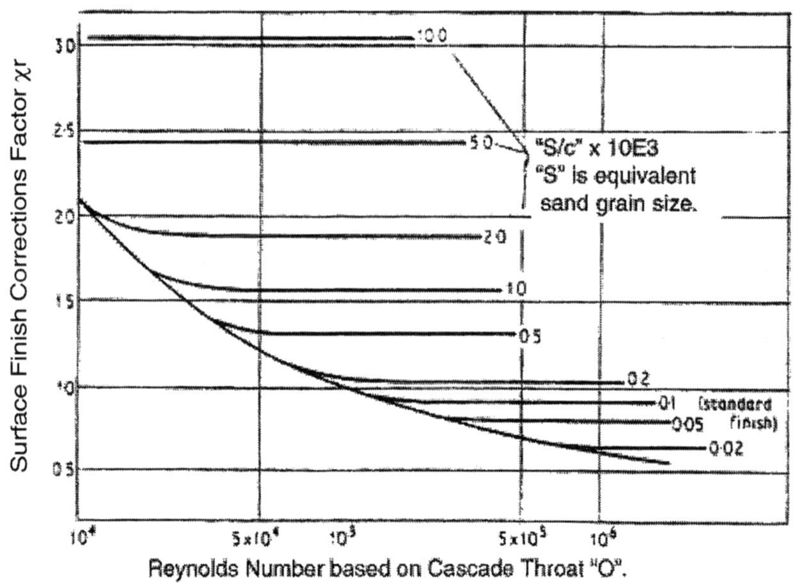

Fig. 3–29 Surface Finish Correction Factor to the Basic Profile Loss to Correct for Surface Finish

In the method of Traupel, the Reynolds Number is calculated based on the profile chord *C*, and the losses are established in terms of a ratio of surface finish *ks* to this same chord *C*. However, in the method of Craig and Cox, the physical characteristic of the profile, the skeleton line length *Cf* is used to determine the significant ratio of surface finish to vane parameters, but the Reynolds Number is determined based on the cascade dimension of throat *O*.

The method of Forster. The method of Forster employs a similar definition of surface finish to that given by Craig and Cox above. It is the equivalent sand grain roughness. However, the dimensional ratio used to define the row geometry is that of *ks/B*, where *ks* is the sand grain roughness and *B* is the cascade axial width.

The method of Forster states that the total drag coefficient ψd is the sum of the form drag ψf and the friction drag ψs. That is:

$$\Psi d = \Psi f + \Psi s \tag{3.26}$$

where

ψd is total drag coefficient

ψf is the form drag coefficient

ψs is the surface finish drag coefficient

For turbulent flow, the suggested value of the surface drag coefficient ψs is from: (Prandtl and Schlichting 1934)

$$\Psi s = \frac{0.455}{\log (Re)^{2.58}} \tag{3.27}$$

For fully developed rough flow:

$$\Psi s = \left[1.89 + 1.62 \log \frac{B}{ks} \right]^{-2.5} \tag{3.28}$$

Further, Forster provides two curves of friction coefficient for the passage plotted to a base Reynolds number. The first curve, Figure 3–30, is for a stationary blade and Figure 3–31 is for a rotating blade. Since these curves were derived considering an impulse type arrangement, they might be more applicable to a definition such as Figure 3–30 as applicable to a row of high reaction and low turning angle. Figure 3–31 might be more applicable to a row with low reaction and a high turning angle. It should be noted that the curves of Forster are corrections to the velocity coefficient ψd and not the efficiency. These curves represent direct losses in the blade rows.

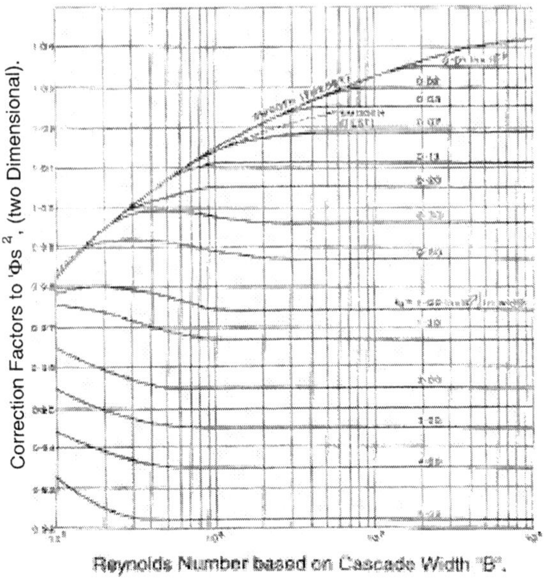

Fig. 3–30 Surface Finish Correction for Stationary Blades

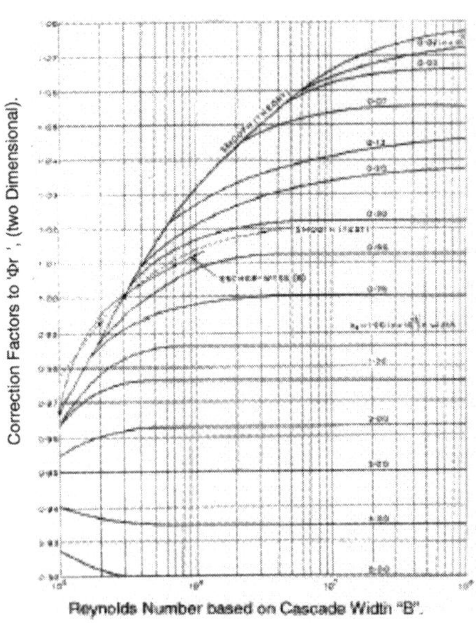

Fig. 3–31 Surface Finish Correction for Rotating Blades

On curves in Figures 3–30 and 3–31, the ordinate is defined as ψp the correction factor on the basic velocity coefficient. The value of ψp is determined from equation 3.29.

$$\psi p = 1 - (\psi f + \psi s)\frac{S\mathrm{þ}}{O}\left[(1-\lambda) + \sqrt{1-\lambda}\right] \quad (3.29)$$

where

Sþ = the passage length as defined in Figure 3–32 for both forms of profile described by Figures 3–30 and 3–31

O = the discharge throat or opening

and λ = $1 - (\sin\beta 2/\sin\beta 1)^2$
which is equivalent to the ratio of the inlet to discharge velocities of the row

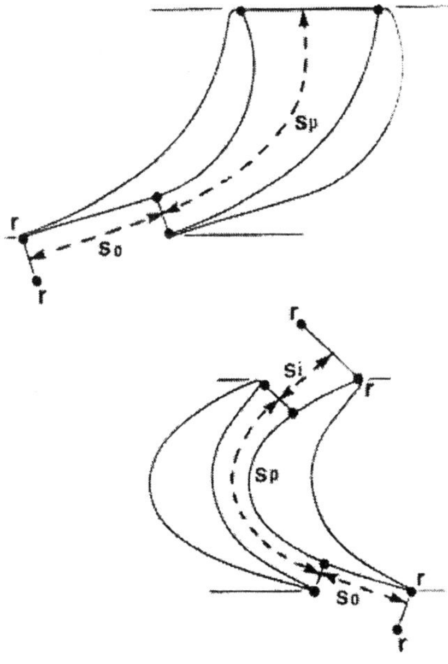

Fig. 3–32 Developed Passage Length for Typical Profiles

Shown in Figure 3–32 are the developed passage lengths *S* as defined by Forster for the high reaction low turning angle cascade *a,* and the low reaction large turning angle cascade *b*. In both cascades the portion of the expansion passage is shown as *Sp,* the outlet portion as *So,* and in the high turning angle cascade the inlet portion as *Si*. At the inlet and outlet positions, the passage lengths *Si* and *So* have been extended until the flow is remove from the effects of adjacent profiles to positions *r-r*.

Shown as Figure 3–33 are the typical requirements for surface finish of two units. In *a* are the requirements for an impulse design, and in *b* for a reaction unit of the same rating. These values should be considered only as typical.

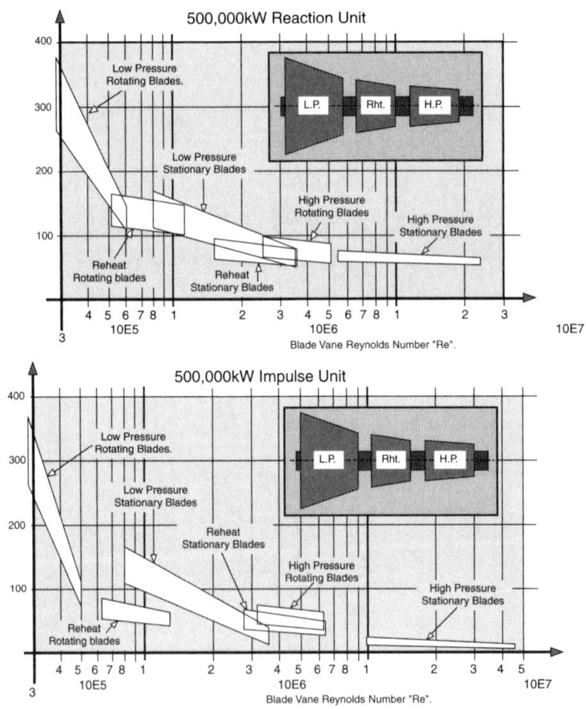

Fig. 3–33 The Typical Range of Surface Finish Requirements in the Blade Rows for Both Impulse a and Reaction b Large Rated Units

Stage discharge Mach Number *(χm, $\chi \Delta m$)*

The stage discharge Mach Number is the ratio of the relative velocity of the steam to the critical velocity in the local steam environment. (The critical velocity is the velocity of sound, which is the speed at which pressure impulses are transmitted through the working medium.) When local velocities in passages formed by two profiles do not exceed the critical velocity. They are subsonic and the variation of pressure in the passage and the region immediately downstream of it is continuous, predictable, and exhibits no serious discontinuities. Under conditions of subsonic velocities, the working fluid changes pressure gradually and pressure and velocity changes occur without shock. (Abbot and von Doenhoff 1958) (Prandtl and Schlichting 1934)

At supersonic velocities, no pressure impulses from the profiles can be transmitted ahead. Therefore, pressure and velocity tend to stagnate or remain relatively unchanged along the profile until the influence of the profile is removed from the working fluid. In the blade passage, there is a relatively minor change of pressure and velocity. However, major changes occur in both at the discharge point and occur over an infinitely short distance with the generation of a shock wave. Shown in Figure 3–34 is a Schlieren photograph of shock waves forming at the discharge point on the trailing edge of a blade profile in a test arrangement. Shown in Figure 3–35a and b are plots of the mach numbers around the tip and root sections of a rotating blade designed for critical velocities. (Forster 1966–67) These plots show the generated shock waves or pressure discontinuities at the discharge portion. In the root section of the rotating blades,

Fig. 3–34 A Schlieren Photograph Showing the Flow Patterns Around a Blade Cascade

where velocities and stage pressure drops are smaller, the difference in the extent and magnitude of the shock waves when compared to the tip section can be clearly seen.

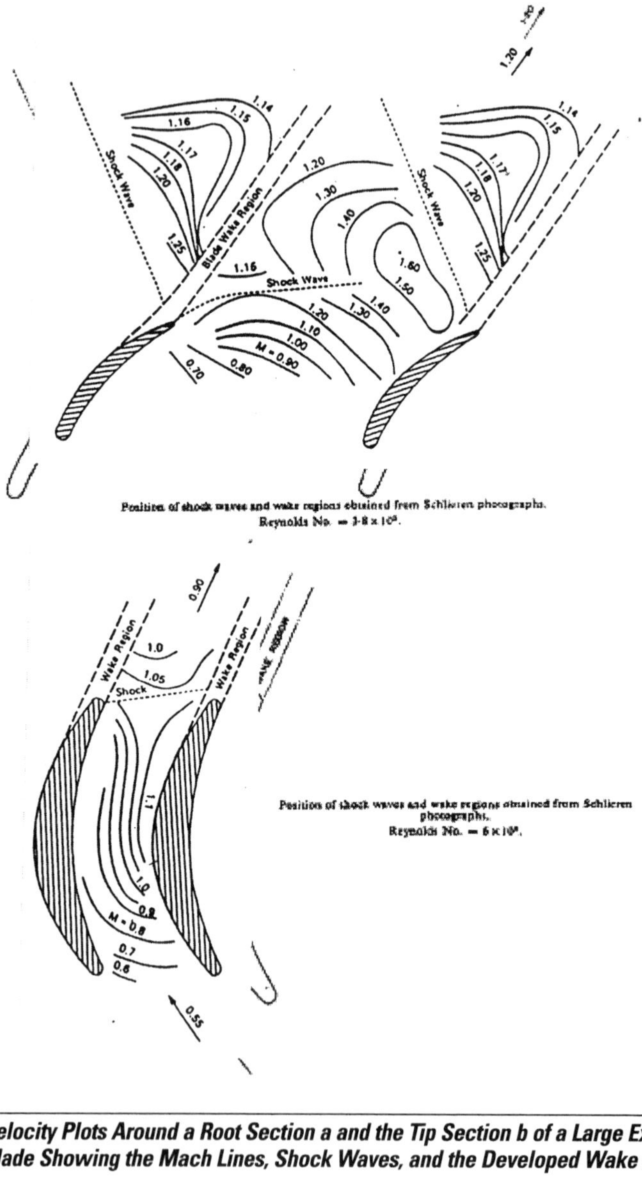

Fig. 3–35 Velocity Plots Around a Root Section a and the Tip Section b of a Large Exhaust Rotating Blade Showing the Mach Lines, Shock Waves, and the Developed Wake Region

Correction factors for the discharge Mach Number *M2* have been established by various researchers. Principle among these being:

The method of Craig and Cox. The method of Craig and Cox provides a correction curve, shown as Figure 3–36, that allows a correction χm to be determined in terms of the isentropic mach number *M2i*. This curve is shown for different values of the actual discharge angle from the cascade, as developed in chapter 1. These curves are shown for different values of the actual discharge angle $\alpha 2$ or $\beta 2$, from the cascade.

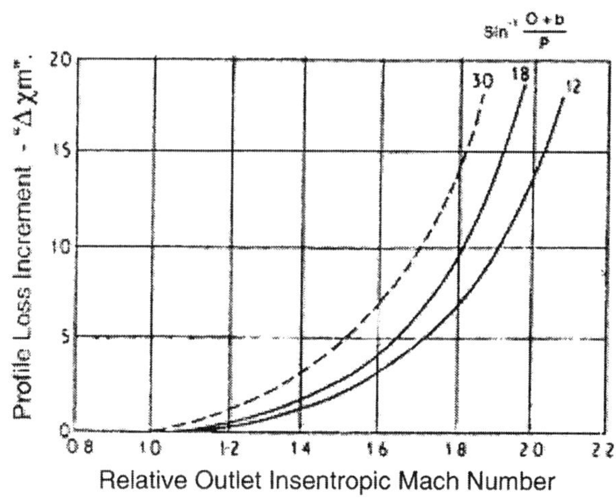

Fig. 3–36 Correction Factor for Discharge Mach Number

The method of Chen. The work of Chen has developed a series of curves which are design specific, and therefore do not lend themselves to general application. These correction factors are dependent upon the profile turning angle θ but do provide for each profile a suitable correction factor ζm. One series of such curves are shown in Figure 3–37 for profiles with turning angles from 15° to 75°.

The method of Chen provides the correction curve as shown as Figure 3–37. The figure shows a correction *ζm* for typical sections at the root, mean, and tip of a blade as a function of the isentropic Mach Number *M2i*. These data assume the steam enters the blade row with zero incidence. The geometric parameters contained in Table 3–1 are applicable to the tested sections.

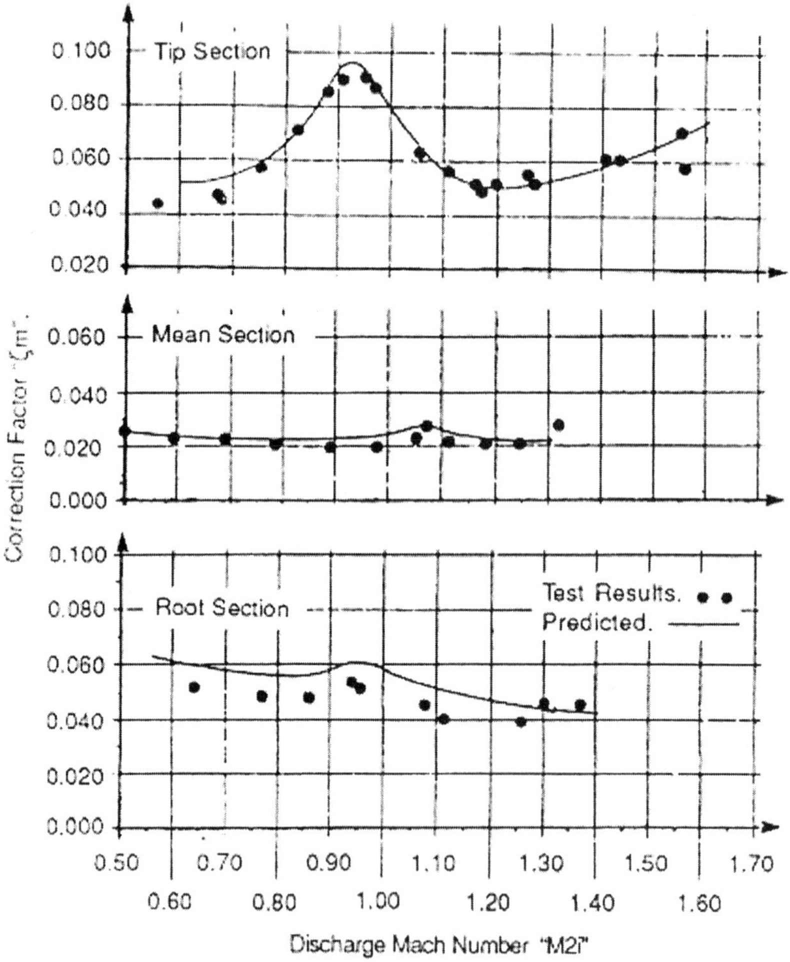

Fig. 3–37 Correction Factor for the Discharge Mach Number

From these curves, it can be seen that losses increase as the discharge velocity approaches the transition zone—as the velocities approach a Mach Number of 1.0. When assessing the losses in large radial blades it is obvious that because of the vortex effect and the variation of the isentropic enthalpy drop across the vanes the Mach Numbers will vary with radial height and also, therefore, the loss factors.

Section:	Symbol	Root	Mean	Tip
Profile inlet angle	$\alpha 1$ or $\beta 1$	152.0	67.0	49.0
Passage discharge angle	$\beta 2$	0.000	5.960	0.000
Profile pitch (mm)	P	23.85	49.78	9.754
Passage discharge throat (mm)	O	6.909	16.740	6.636
Skeleton line length (mm)	Cf	35.89	100.80	42.29
Trailing edge thickness (mm)	b	0.508	1.755	0.305
Surface roughness (mm)	ks	65.50xE-4	63.50xE-4	31.75xE-4

Source: Chen, S. "A Loss Model for the Transonic Flow Low Pressure Steam Turbine Blades."

Table 3–1 Blade Geometric Parameters

Discharge trailing edge (χb, $\Delta\chi b$)

A wake region is formed downstream of the discharge edge, and such regions can be seen in the previous velocity plots of Figure 3–35a and b. These plots have been determined experimentally and are correct for these profiles at the pitch and setting angles shown. These plots will also be influenced by the environmental conditions under which they are operating. It is possible that these or similar profiles, with only minor modifications to the cascade or profile trailing (discharge) edge, would exhibit somewhat different wake regions while operating under the same or similar environmental conditions. Different wakes would also be obtained with marginally different pitching. The wake that is formed at, or close to, the point on the suction face of the profile where the boundary layer separates is influenced by the curvature of the suction face in this region is discussed in chapter 8.

Often, but not invariably, this point of flow separation from the suction face is coincident with or very close to the point in the passage at which the throat or discharge opening occurs. At the separation point, vortices are formed and shed from the surface. From this separation point, the vortices are transported downstream with the working fluid, and any region between the boundary of these vortices and the surface from which they are shed will be turbulent and contain high-energy vortices. The other extreme of the wake region is from the discharge point on the pressure face of the profile. There is therefore considerable advantage to making the trailing edge as thin as possible. However, such thinning must be done recognizing there are limits to which the discharge can be thinned.

- There is a possibility of material loss on both stationary and rotating elements through erosive mechanisms, particularly in stages after the entry or return of steam from the boiler superheater and reheater sections. This loss would destroy the throat position and discharge area and angle
- As the trailing edge becomes thin on rotating blades, there is a possibility of local stresses reaching values in which there will be material rupture.

There are various forms of trailing edge produced on both stationary and rotating blades. This trailing edge shape is influenced to a certain degree by the process used to determine and manufacture the blade. A trailing edge with a rounded suction face as shown in Figure 3–38a rather than a flat as seen in Figure 3–38b will tend to increase the wake thickness to *T*. If the discharge point is rounded, as shown in the inset to Figure 3–38a, this will tend to introduce some flow following the pressure surface around the discharge point, causing separation to occur at point *b*. This effect will reduce the wake thickness by some small amount *dt*. This reduction will obviously depend upon the trailing edge radius, the surface roughness, and direction of finish on the trailing edge portion of the vane.

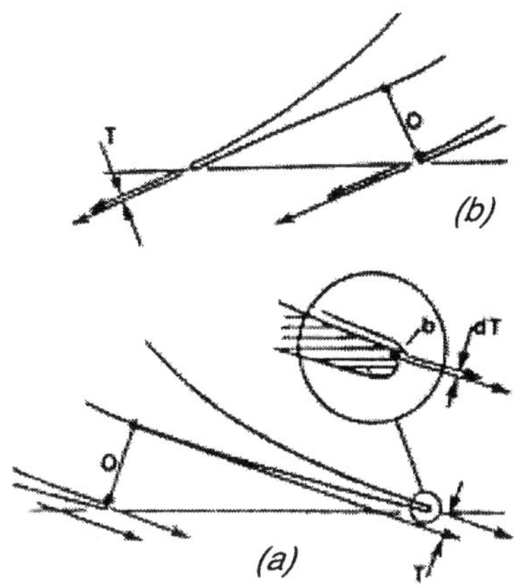

Fig. 3–38 Details of Trailing Edge Forms, and Discharge Tail Thickness

The wake region has been discussed in two-dimensional terms. In fact, the wake region can and normally does exist along the radial length of the blade vane and will consist of a vortex sheet. This vortex sheet is unstable and will attempt to roll up into a simple vortex as soon as the surrounding environment permits. Normally, for a turbine stage such roll up is unlikely to occur as there is insufficient axial space. However, for the large axial gaps that often exist at one end of large double-flow low-pressure sections, this role up could begin at some radial positions, particularly at light load.

A correction factor χb has been established for the trailing edge thickness by Craig and Cox.

The method of Craig & Cox. The basic profile loss, by the method of Craig and Cox assumes a trailing edge thickness of zero, as noted earlier in this chapter. Therefore, because of the potential for losses associated with the trailing edge, it becomes necessary to have a correction factor which modifies this basic loss factor ζbl for a vane of finite thickness b. This correction factor, presented by Craig and Cox as χb, is shown in Figure 3–39, which shows values of χb to a base of trailing edge thickness to pitch for various steam discharge angles $\beta 2$. This discharge angle is the steam angle determined from the vector diagram. This angle $\beta 2$ is distinct from the metal section angle $\beta 2o$. Also shown on this curve are the additional losses $\Delta\chi b$ as required in equation 3.21.

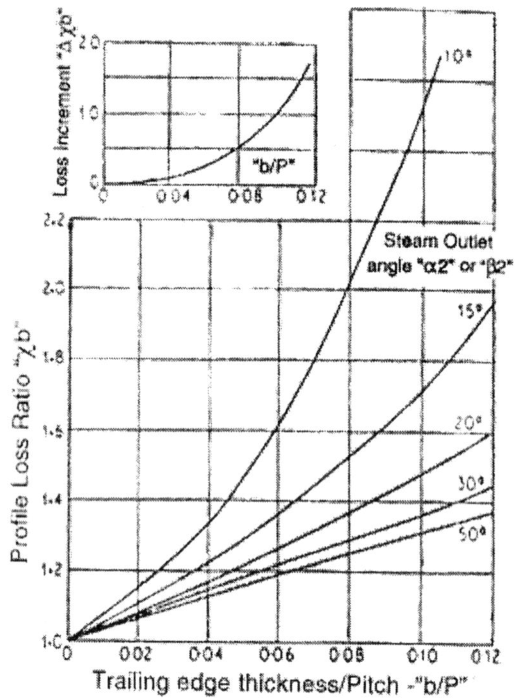

Fig. 3–39 Trailing Edge Thickness Correction

This same correction factor is also recommended for use by Chen when employing his method.

Methods described by Horlock. Horlock describes various methods of establishing the losses associated with the thickness of the trailing edge. These include:

• The Method of Markov: The method of Markov uses an expression that provides a correction factor χb for an edge thickness b to some known loss at a thickness bo with a loss factor χbm. (Markov 1968) This expression of Markov is

$$\chi b = \chi bm + 0.41 \left[\frac{b - bo}{P. \cos (90-\beta 2)} \right] \tag{3.30}$$

where

χbm = known loss at thickness bo
χb = predicted loss at thickness b
b = trailing edge thickness for which loss is required
bo = trailing edge thickness for which loss is known
P = vane pitch
$\beta 2$ = steam discharge angle

In the case of a loss assuming no losses at a trailing edge thickness of zero $bo = 0$, which is used in many profile loss predictions, it is then possible to predict the losses when the trailing edge has a thickness of b. In this case, equation 3.30 would be reduced to:

$$\chi b = 0.41 \left[\frac{b}{P. \cos (90 - \beta 2)} \right] \tag{3.31}$$

• The Method of Ainley: This method provides an approximate expression for thickness loss as follows:

$$\chi b = \chi bm' \left[1 + 7 (b - 0.02) \right] \tag{3.32}$$

where

χbm = known loss at thickness at ratio of b/P of 0.020

This expression is shown for a ratio of b/P = 0.02 in Figure 3–40.

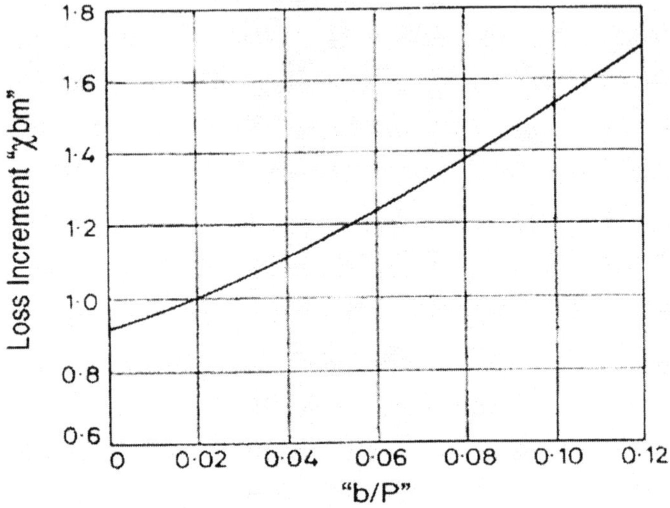

Fig. 3–40 Training Edge Thickness Correction as a Function of the Ratio b/P

Suction surface trailing edge curvature *($\Delta\chi c$, $\Delta\chi s$)*

Depending upon the method of determining the profile for any row, the trailing edge suction surface can be of a form other than straight. In this case, the trailing edge is defined as that portion of the suction surface from the throat to discharge edge. If the throat is formed on the profile trailing edge, which has curvature as shown in Figure 3–38a, a turbulent region is produced between the steam stream flow and the profile suction face. This region is one where vortices are formed that are eventually shed from the surface, and travel downstream to the following blade row with the main steam flow. This high energy, turbulent region is known as a wake region. In the case of a straight- or almost straight-backed suction face as in Figure 3–38b, this region either does not exist to any significant extent or is relatively thin on the suction face location. Developed wake regions are shown in Figures 3–35a and b.

The angle of deviation, as defined in Figure 3–41, is the difference between the metal section angle $\beta 2o$ and the discharge angle $\beta 2$ determined by the equation relating the opening O to the pitch P. These angles determine the angle of deviation Γ and can be established from equation 3.33. The actual discharge angle of the steam is more nearly:

$$\Gamma = \beta 2 - \beta 2o \tag{3.33}$$

The actual discharge angle of the steam is more nearly $sin^{-1}\ O/P$.

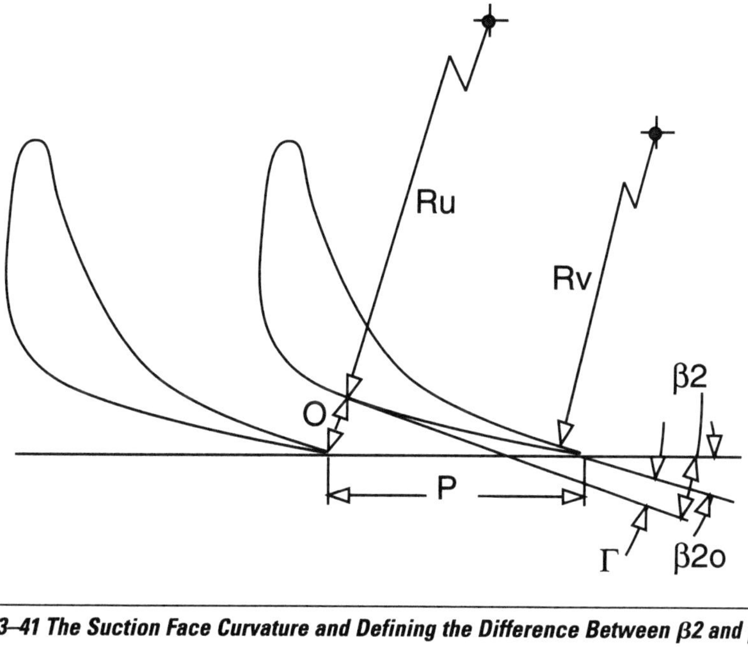

Fig. 3–41 The Suction Face Curvature and Defining the Difference Between β2 and β2o

Methods accounting for the losses introduced by these wakes have been formulated. Two of these, in terms of a correction factor $\Delta\chi c$, follow.

The method of Chen. The work of Chen relates to velocities at about the mach 0.9 to 1.2 where losses are due to shock and boundary layer interaction. Therefore, the correction factors he determined are not appropriate in all cases.

Shown as Figure 3–41 is the basic geometry of the training edge for a profile with curvature on the suction face in the region of discharge and throat formation. The curvature is *Ru* at the throat, *Rv* at the discharge point, and the average curvature is *Ra*. This curvature can be the same along the entire length of the trailing edge or vary, depending upon the method of establishing the trailing edge form.

Chen states the relationship between the mean curvature *Ra*, and the angle of deviation Γ is given by equation 3.34.

$$\frac{P}{Ra} = \frac{\text{Sin}\,\Gamma}{1/2\left[\sqrt{\{(O/P)^2 \,.\, \text{Sin}^2\,\Gamma + 2[1 - (O/P)^2] \,.\, [1 + \text{Cos}\,\Gamma)\}}\right] - 1/2[O/P \,.\, \text{Sin}\,\Gamma]} \tag{3.34}$$

The loss χc is a function of the ratio *P/Ra* and the discharge mach number.

The method of Craig and Cox. The method of Craig and Cox provides a loss correction curve. Figure 3–42 shows the loss as a function of the ratio pitch *P* to suction face mean radius *Ra*. It can be seen from this correction factor curve that the magnitude of losses increases significantly as the mach number increases, and for many stages of low discharge velocity there is no need to make any correction for losses.

As a guide, profiles should not be used at even the lower mach numbers at a setting angle or pitch that will make Γ exceed 7°. This is often difficult in cylindrical or constant profile blades where towards the tip section the pitch has increased to the extent the throat is formed on the curved portion of the vane.

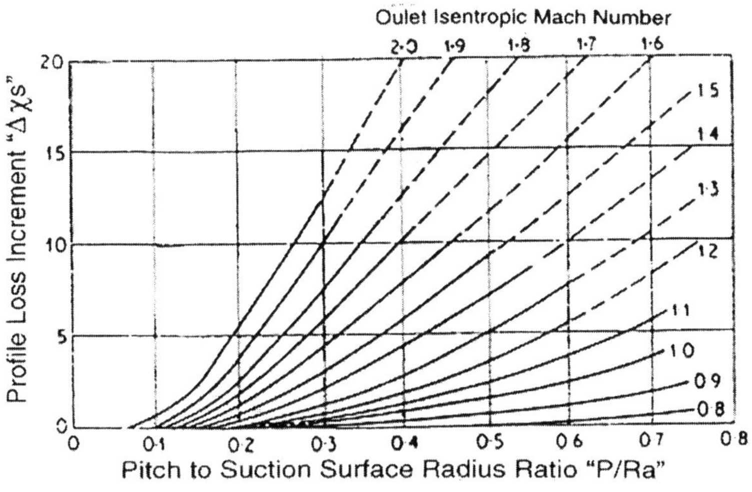

Fig. 3–42 Blade Suction Surface Radius Losses as a Function of Pitch P to Suction Face Radius RA

Incidence and the inlet nose *(χi)*

The blade profile is designed to accept steam entering at an angle predetermined by thermal calculations. There is little apparent reason why differences should exist between the steam and profile metal angles. In fact, however, such differences often exist. Principal reasons for these discrepancies between design and actual follow.

- As unit load varies, steam conditions and flow quantities alter the pressure distribution and steam velocities on certain stages (control and exhaust), and this causes minor changes of direction in steam flow angles. The remaining stages are designed and operate with a constant pressure ratio. Therefore, the inlet angles to the rows do not change significantly.

- The designer will often elect to use existing profiles of proven performance and known characteristics rather than design a new element requiring expensive and time-consuming development.

- Flow considerations of many profile forms will often allow a relatively blunt inlet nose that, being rounded, will allow their use and not incur excessive losses even if the metal angle does not match the steam inflow angle. This will require the steam deflect around the nose to flow over the pressure or suction faces.
- Manufacturing inconsistencies and blade setting angles can cause minor errors in the vane inlet angles.

Consider the form of the inlet nose shown in Figure 3–43. As the steam approaches the inlet, it must deflect to pass over either the pressure or suction face. This deflection leaves a small zone *S* just ahead of the inlet nose in which steam velocity is destroyed. This is known as the stagnation zone. Here the destruction and the reconversion of kinetic to thermal potential energy produce a rise in steam and vane metal temperatures.

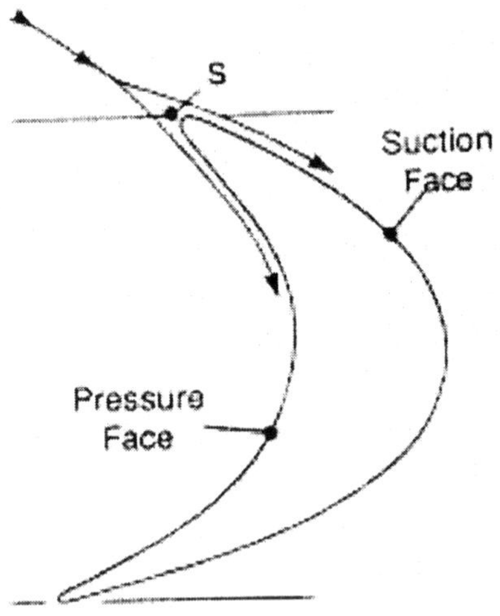

Fig. 3–43 Inflowing Steam and Stagnation Point S

The losses that occur as a consequence of the incidence between the profile metal angle and the steam inlet angle are a function of several variables. These include the form of the inlet nose, whether the inlet nose is radiused, and the value of that radius. The loss will also be influenced by the Reynolds Number at inlet and the surface finish. If the nose is produced by a hand-polishing operation, common in many manufacturing techniques, then the manner in which the radius conforms to the design specified requirement will influence the incidence angle.

In examining profiles, it should be noted that the shape of the inlet nose is influenced considerably by the turning angle of the profile. For large turning angles, a smaller radius is required to allow sufficient inflow area for the steam. For smaller turning angle vanes, it is possible to employ a larger radius as there is more tangential space available.

There are correction curves available, and the following are considered typical.

The method of Craig & Cox. Craig and Cox have provided a curve that provides an incidence loss ratio χi in terms of an incidence parameter Ri, which is defined as:

$$Ri = \nu - \frac{\nu min}{\nu stall - \nu min} \tag{3.35}$$

where

ν = the incidence angle in degrees

νmin = the incidence for minimum loss

$\nu stall$ = the stalling incidence

The parameter χi is shown plotted to a base of Ri in Figure 3–44.

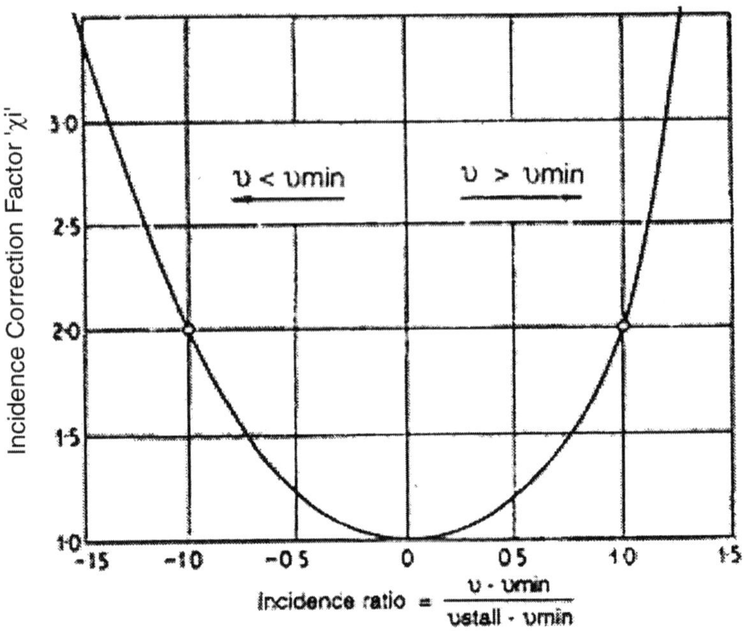

Fig. 3–44 Incidence Correction Factor

The method of Chen. The method of Chen provides incidence loss factors directly in terms of the incidence angle for various profile metal inlet angles $\alpha 1$ and $\beta 1$. These are shown in Figure 3–45a for inlet angles from 35°to 90° and in Figure 3–45b for inlet angles from 90° to 135°.

The methods described by Horlock. Horlock examines the results of several researchers and provides curves showing the losses as a function of the incidence angle. Figure 3–46 shows comparative results of Soderberg and Ainley, for different ratios of thickness/chord ratio but considers the data of Soderberg to be somewhat over simplified. Better data is available from Ainley as shown in Figure 3–47, which is typical.

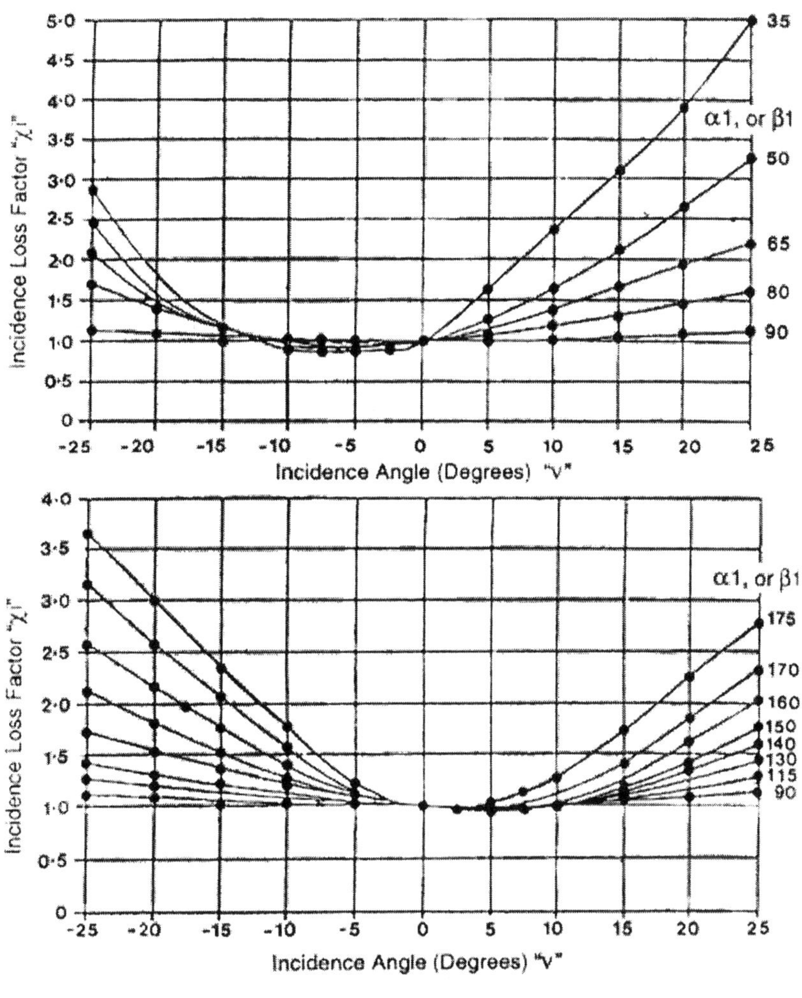

Fig. 3–45 Incidence Loss Factors for Inlet Angles in a from 35 to 90° and in b from 90 to 175°

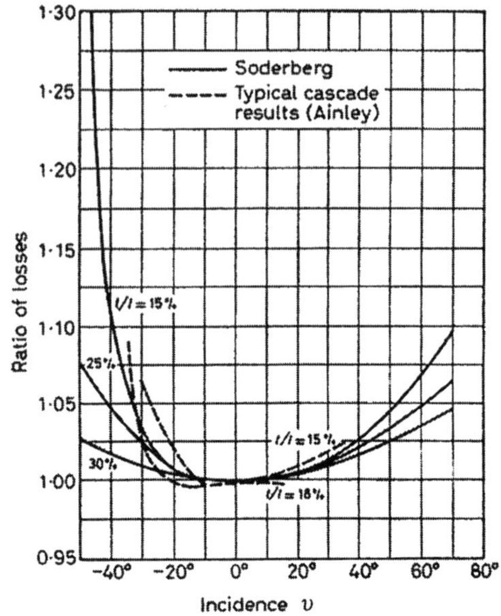

Fig. 3–46 The Test Results of Soderberg and Ainley Shown as a Loss Coefficient

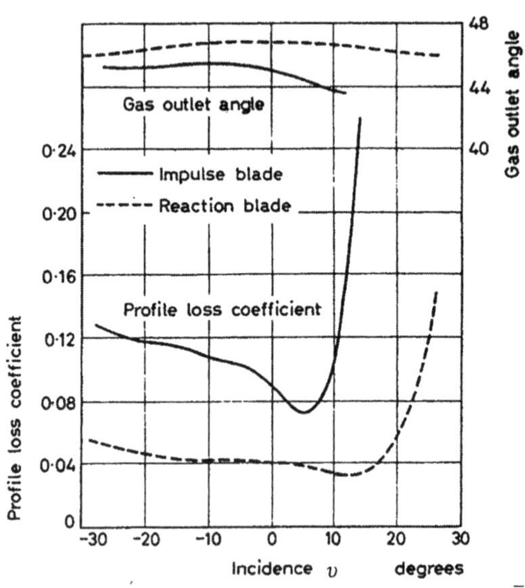

Fig. 3–47 The Losses of Soderberg and Ainley Reduced to a Loss Ratio

TIE OR LACING WIRE LOSSES

Tie wires are included in many long blades to help dampen the effects of blade vibration by transmitting vibratory motion from impulses developed in one element to another connected by wires and within the same blade group. Unfortunately, these wires will be the source of considerable energy losses due to the drag effect of the wire and the disturbances they introduce into the steam flow in the blade passage. Shown in Figure 3–48 are the deposit patterns behind a strengthening stub in a stationary blade row—in this case placed to add security to a single stationary blade vane potion, but the inference of flow disturbance is clear.

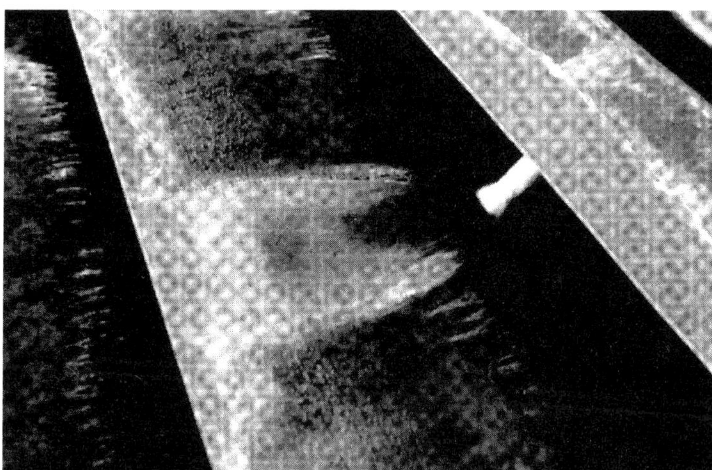

Fig. 3–48 Showing the Flow Separation Around a Wire or Stub in the Steam Path

The effects of losses due to the tie wires can be determined using the following methods.

The method of Craig & Cox. This method requires knowledge of the steam properties at the axial location of the wire and provides the following formulation for determining the magnitude of the loss introduced by the wire.

$$\Delta \eta t = \left[\frac{\frac{Aw \, . \, Cd}{Ad} \times \frac{Ww^2}{2gJ}}{Pb} \right] . \, \eta b \qquad (3.36)$$

where

$\Delta\eta t$ = the wire energy loss

Aw = the wire cross sectional area

Cd = the wire drag coefficient

Ad = the passage flow area at the wire axial location

Ww = the steam stream velocity at the wire axial location

Pb = the work done by the blading

ηb = the blading efficiency

The drag coefficient can be taken to be *Cd=1.0* for circular wires and suggests that for elliptical wires this value can be reduced to 70% if the major-to-minor axis of the elliptical wire is in the ratio 1 to 4.

STAGE GEOMETRIC FACTORS GROUP TWO LOSSES

The losses discussed earlier are those related to the form and arrangement of the profiles and those associated with the tie wire, which in effect is an obstruction included in the blade passage for mechanical reasons. As such, these losses are predictable with a reasonable degree of accuracy in undamaged profiles, and the variations that occur as a consequence of changes in steam conditions and flow angles can also be predicted and accommodated with reasonable accuracy.

The steam parameters that exist along the radial height of the blade are changing due to the radial flow or vortex effect. There is some advantage to the use of vortex or twisted vanes that are adjusted to more nearly meet the requirements of the flowing fluid. Cylindrical (constant section) vanes will not meet these requirements as closely as the vortex element because the steam properties will continue to change in the radial direction. There will also be increases in vane pitch with increasing diameter making the losses even greater if the vane profiles are not changing to meet steam requirements. These losses can also be predicted with a level of accuracy sufficient to allow stage losses to be determined.

The major group two losses can then be determined from the following.

Disc friction *(χw)*

When the rotating blades are carried on wheels or discs, these wheels are immersed in the working fluid that has no design specified direction but simply fills the space between the diaphragm and wheel on both sides. The relative motion between the wheels and steam causes a frictional loss to be developed. The magnitude of this loss is a function of the steam density at the wheel position and the amount of exposed area.

There are a number of formulations for determining these losses, and they were summarized by Kearton based on relatively early work. (Kearton 1958) These include:

The method of Stodola. As the result of development work in air, Stodola developed the expression (Stodola 1924):

$$hpf = \frac{C}{10^6} \cdot D^2 \cdot u^3 \cdot \rho \tag{3.37}$$

where

- hpf is the horse power lost to disc friction
- C is a constant = 0.071 for air
- D is the effective diameter of the disc in feet
- u is the tangential velocity of the disc in feet per second
- ρ is density of the working fluid in lb per cu. foot

The method of Ribrary. A similar formulation developed by Ribrary follows. (Ribrary 1934)

$$hpf = C \cdot D^2 \cdot \left[\frac{u}{100}\right]^3 \cdot \rho \qquad (3.38)$$

Where

$$C = 1.90 - 0.2.\log\left[\frac{D \cdot u \cdot \rho}{\mu}\right] \qquad (3.39)$$

μ is the absolute viscosity

The method of Craig & Cox. The method of Craig and Cox, again provides a formulation which allows these losses to be determined. The expression given is:

$$\Delta\eta d = \left[\frac{Pl}{Pb}\right] \cdot \eta b \qquad (3.40)$$

where

- Δηd is the reduction in stage efficiency due to windage
- Pl is the power loss due to windage given by equation 3.41
- Pb is the work done by the blading
- ηb is the blading efficiency

$$PI = Tc \cdot \left[\frac{\left[\frac{RPM}{100}\right]^3 \cdot Dh^5}{m \cdot Vs} \right] x \frac{10}{3.471^{-10}} \text{ BTU/lb} \qquad (3.41)$$

where

Tc	is	the torque coefficient
RPM	is	the rotor speed in rpm
Dh	is	the rotor hub diameter
M	is	the flow of steam through the blade annulus
Vs	is	the steam specific volume

The torque coefficient *Tc* is a function of the geometry of the stage and also of the Reynolds Number for the disc. The Reynolds Number is influenced by the surface finish of the disc, which in a new unit can be controlled by design specification. However, as the unit ages, there can be a tendency for this surface to deteriorate. Figure 3–49 shows is a disc that was originally finished by design specification to 125√μ-inches, but that has deteriorated to the level 1000-1500√μ-inches because of moisture impingement on the vertical faces. It can be seen that the harder pins attaching the blade to the wheel have not been affected by this moisture.

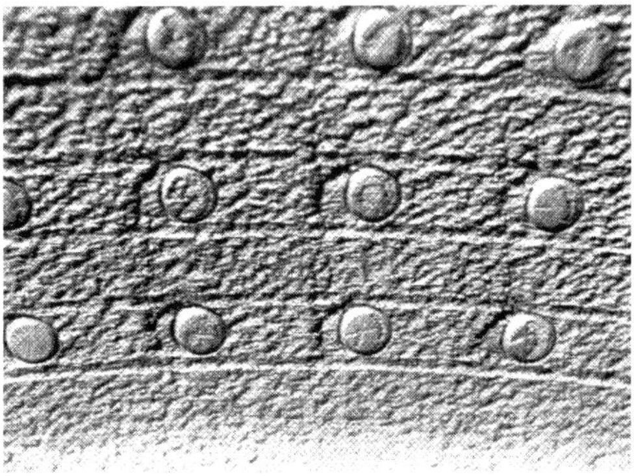

Fig. 3–49 Water Roughening of a Disc

Row aspect ratio *(χar)*

As steam expands (reaction) or flows without pressure drop (impulse) through a blade passage, it changes direction and there is a frictional interaction between the steam and the inner and outer sidewalls. This is a result of both the steam velocity and its change of direction. The resulting effect of this frictional loss is that at the inner and outer diameters of the passage, there are developed vortices that are shed from the inner and outer regions of the expansion passage as shown in Figure 3–50. These represent losses to the blade row.

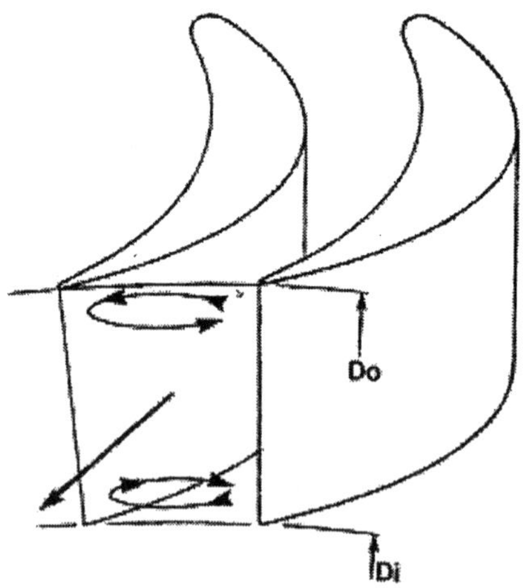

Fig. 3–50 Flow Circulation at the Inner and Outer Steam Path Boundaries

These vortices occupy a portion of the total discharge area, and as the vane height increases, this portion represents a smaller portion of the total area. It would be expected that with large radial height blades this loss become less significant but still present.

Various definitions have been used by experimenters to define this aspect ratio loss. For a turbine blade row, the aspect ratio *Ar* is defined in terms of the blade radial height *h* to the chord *C*, the axial width *B*, or the skeleton line length *Cf*. Therefore, in interpreting any data related to aspect ratio, clear definition of the ratio involved should be established. The following represent typical loss coefficients.

The geometric parameters of a blade row are a function of steam flow quantities and stress requirements. Therefore, the designer has very little, or at least limited, choice in their values. Also, aspect ratio losses are relatively minor in all stages, except the first and possibly second in any section where the influence of inner and outer sidewalls can be relatively large due to secondary flow effects influencing a larger proportion of the total discharge height.

The method of Craig & Cox. The aspect ratio, as defined by Craig and Cox is the ratio of the vane radial height *h* to the profile skeleton line length *Cf*. This method provides a loss correction factor χar as a function of the inverse of the aspect ratio. This factor is shown as Figure 3–51. As the radial height of the blade increases, the losses become less significant in terms of the overall losses in the stage.

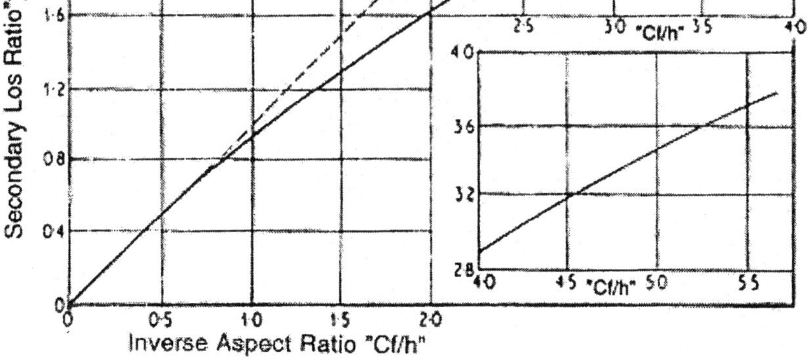

Fig. 3–51 Losses Due to Aspect Ratio and Secondary Circulation

The methods described by Horlock. Horlock describes the aspect ratio as the ratio of vane height *h* to axial chord *C*, and provides correction factors determined by several experimenters. (Horlock 1966) The results of their work is shown in Figure 3–52, which again indicates that as blade radial height increases it reduces the level of losses incurred in terms of a percentage of the total losses in the stage.

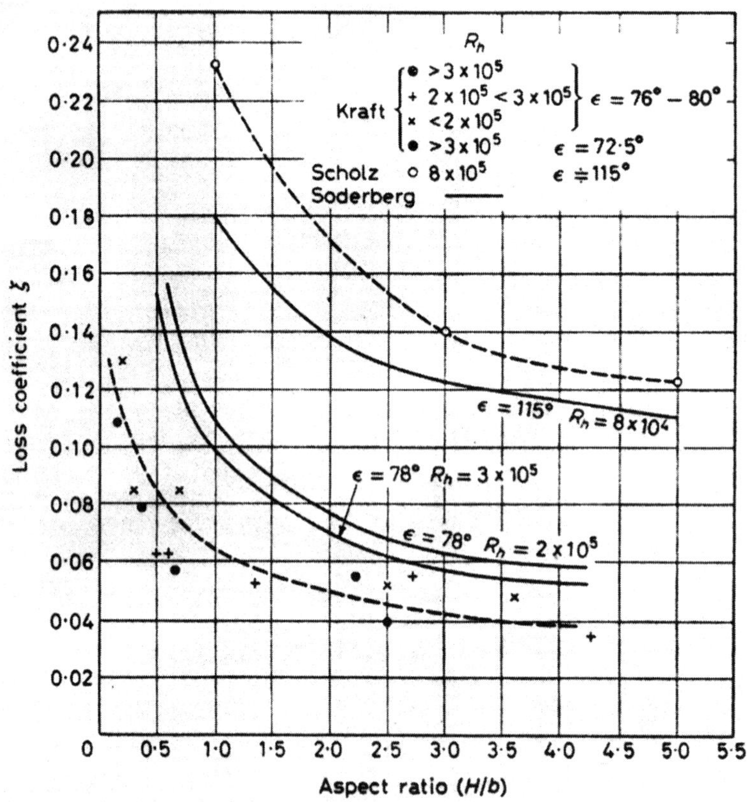

Fig. 3–52 Aspect Ratio Losses

From the curves Figures 3–51 and 3–52, it can be seen that group two losses exist and are of greater concern in those stages of smaller radial height. What is not as obvious from these curves is that they assume the inner and outer sidewalls are smooth, or close to hydraulically smooth, and the loss effect is limited to the vorticity of the steam as it leaves the blade opening. However, there can also be a considerable loss associated with friction on the inner and outer sidewalls, and it is necessary to ensure that these walls are smooth as originally manufactured and are clean and free of deposits and damage when the unit is removed from service for examination and cleaning.

Wall end, flare, and stage transition losses

These forms of losses are conveniently grouped together and refer to those losses that occur at the inner and outer sidewalls of the expansion passage. These losses are related to friction and are influenced by sidewall and surface discontinuities. Normally there are no or minimal discontinuities at the inner sidewall with the minor exception where the vane attaches to the root platform. At the outer sidewall, losses are induced not only by surface finish, but also by discontinuities, many of which are designed into the unit for the removal of feed heating steam and steam for other purposes. This is also seen in stages located in the saturation region for the removal of collected water, or in the case of units designed for automatic extraction, where valve extraction points exist.

These losses are more significant if they occur in the high-pressure stages because of the higher steam density and therefore Reynolds Number. They are less significant in the lower pressure stages but because of the rate of volumetric flow increase, there can be significant coning angles in this region. However, in the low-pressure section of the unit, wall surfaces can deteriorate rapidly under the influence of moisture impact and washing. The choice of casing materials becomes of greater importance here from considerations of performance.

These losses associate with the wall, its geometry, and overall arrangement are examined by Craig and Cox and consider the following geometric parameters.

Annulus wall losses. These losses occur in the steam path due to annulus area increase between stages. They are influenced by the amount of flow area increase from the discharge from one rotating row to inlet to the following stationary row. They are also affected by the axial space between the rows that is available for this area increase to occur. A means of quantifying these losses for different area ratios and different flare or coning angles is shown in Figure 3–53. From these curves, it can be seen that there are corrections given for controlled and uncontrolled extractions. In the case of the uncontrolled expansion, it is necessary to take account of the distance ratio *x/h*, where *x* is the distance between the stages.

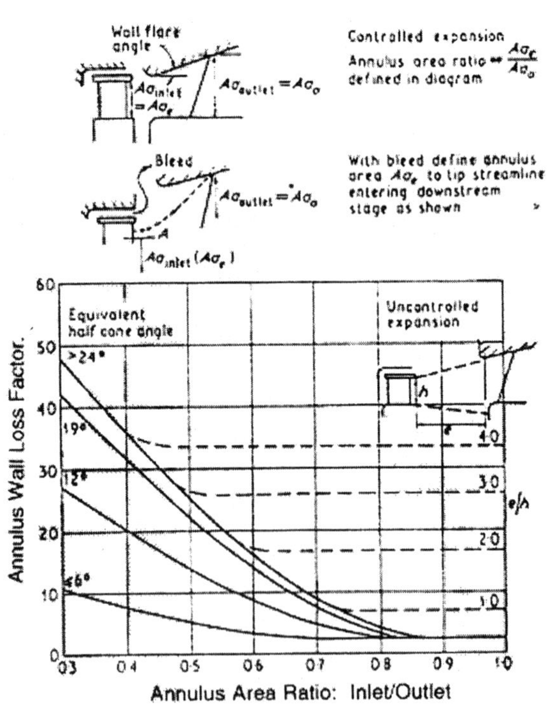

Fig. 3–53 Annulus Wall Loss Correction Factor

Cavity losses. There are reasons for the steam path design to incorporate cavities in the casing inner surface. These cavities are for the extraction or removal of both steam and water from the steam path. The effect of these cavities/discontinuities is to disrupt the orderly flow of steam with the consequence of introducing losses. The magnitude of these losses can be seen from Figure 3–54.

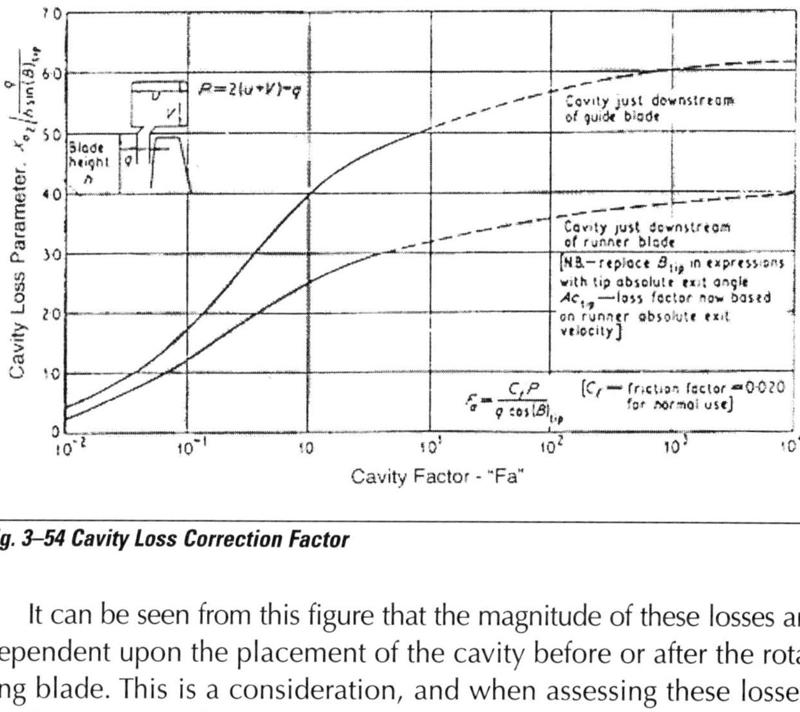

Fig. 3–54 Cavity Loss Correction Factor

It can be seen from this figure that the magnitude of these losses are dependent upon the placement of the cavity before or after the rotating blade. This is a consideration, and when assessing these losses, the location must be accounted for.

Lap losses. The lap, see Figures 1–30 and 1–31, is designed into a stage to promote the orderly flow of steam from one row to the next. It is essential this lap be included in the design to minimize losses associated with discharging steam over the blade tips or into the root platform if it were not included.

Higher losses occur between a rotating row and the following stationary elements in those stages that have a large increase in the steam specific volume, and therefore the requirement of increasing annulus areas from one row to the next.

While it is possible to maintain the inner and outer lap at values that are just sufficient to capture all steam from a discharging row, this sometimes requires the use of a slanted outer boundary, which in the case of some diaphragm manufacturing techniques increases the cost significantly. Therefore, the designer may elect to use parallel sidewalls rather than conical and accept the additional losses.

The losses induced into the stages by lap are shown in Figure 3–55, which is for a rotating stage with a coverband. If no cover is used, then the outer wall is, in effect, the inner surface of the casing, and the distance from the discharging blade row to the casing should be considered as the effective outer lap.

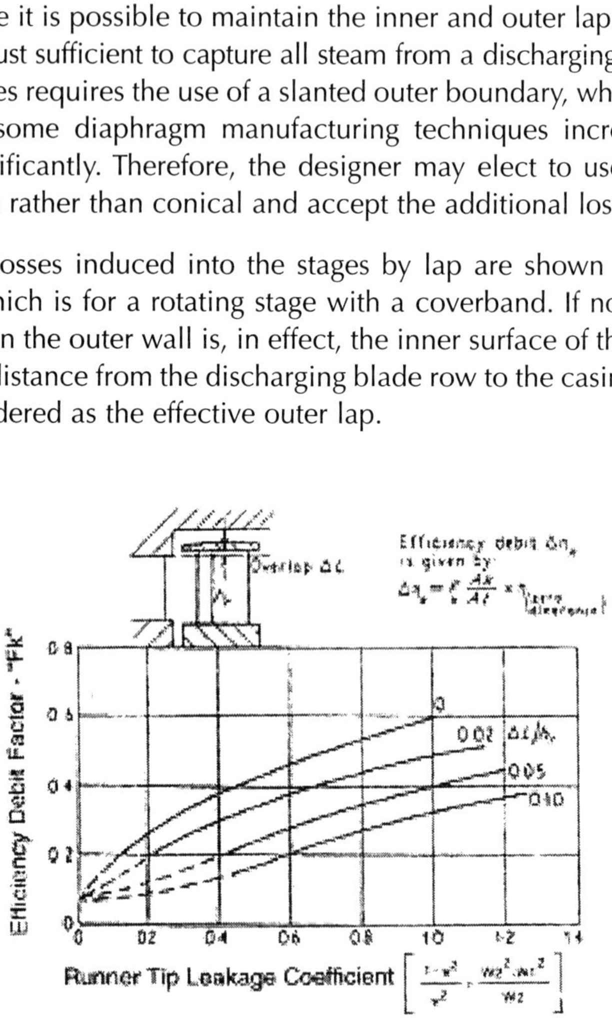

Fig. 3–55 Lap and Coverband Loss

Pressure balance holes

In those stages designed on the impulse and/or low reaction principle where the rotating blade rows are carried on a wheel, either formed integrally with the central spindle or shrunk onto it, it is normal in single-flow sections to include a pressure balance hole in the wheel. This hole is to ensure that any axial thrust that would be developed across the wheel due to pressure differentials, including the effect of mechanical damage, is prevented.

Figure 3–56 shows the typical geometry around a hole, and as there are variations of pressure from *Pi* to *Po,* there will be a flow through the hole to equalize pressure. The normal adjustment, if any is required, is from the inlet pressure *Pi* to the rotating blade row discharge pressure *Po*. However, situations can occur where on the wheel faces the discharge pressure can become marginally higher than the inlet pressure. In either case, there is a modifying flow to adjust the pressures on the wheel face so they are close to the same value, and there is not a differential sufficient to cause excessive net axial thrust.

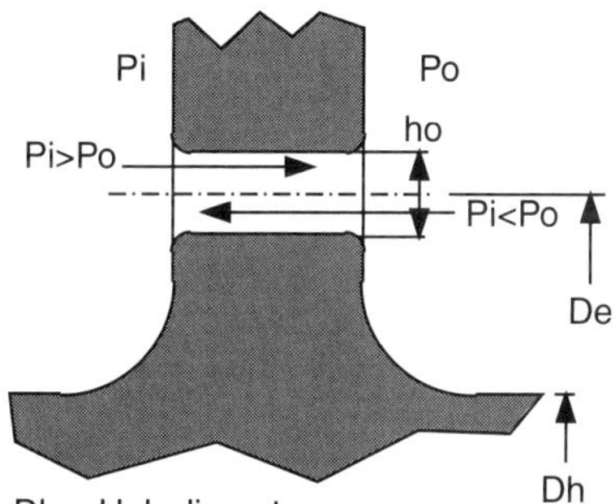

Dh = Hub diameter.
De = Balance hole location diameter.
ho = Pressure balance hole diameter.
Pi = Inlet side steam pressure.
Po = Oulet side steam pressure.

Fig. 3–56 The Pressure Balance Holes in the Rotating Wheel

The leakage flow through these balance holes can be determined by simple expressions. Normally, the number of holes is a prime number, that is five, seven, etc. This is to prevent a resonance. The leakage flow can take one of several leakage paths after passing through the hole. Either it can leak under the following diaphragm or reenter the main steam flow. In the case of leaking under the diaphragm it has a positive sealing effect, but in the case of reentering the main steam flow it will tend to disrupt the main flow causing losses.

Negative root reaction

The reaction (pressure drop) in the rotating blade row increases from the root to tip section as a consequence of the vortex or radial flow component of the steam discharging from the stationary blade row. Circumstances on some stages are such that at the root sections the reaction is negative. That is, at discharge from the rotating blade, the steam has a higher pressure than at inlet. This can occur as a consequence of over-expansion in the stationary blade row, and then in the following rotating row, the steam is recompressed to the stage outlet pressure. While this effect introduces losses, there are some stages that are designed to achieve this condition, so root section recirculation and cooling of the root or blade attachment is induced. Such a pressure rise, although small, will require a pumping action by the blade. This obviously induces losses and lowers stage efficiency.

The possible flow patterns around a stage with negative root reaction are shown in Figure 3–57. This diagram is provided as representative. The important flow is at the rotating blade root, carrying steam from position *R* to *T* and in so doing removing heat from the region of the blade root. The steam flow through the pressure balance hole joins steam that has leaked under the diaphragm from *S*. As shown, pressure *Pmr* between the stationary and rotating blade rows at the stream line is *T*, which is lower than at discharge from the stage *R*—pressure *Po*. This pressure differential encourages backflow of steam through the root section of the rotating blade row or recirculation through the pressure balance hole from *H* to *S*.

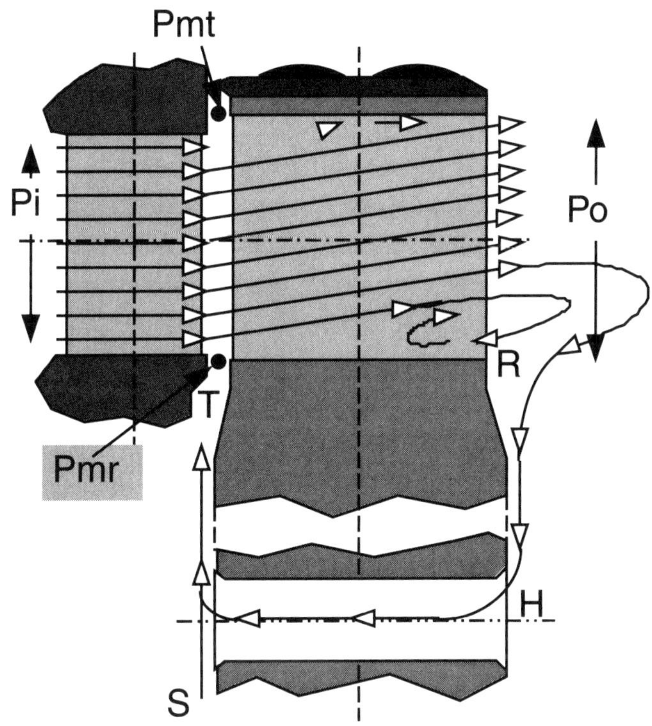

Fig. 3–57 The Effects of Negative Root Reaction

The losses discussed previously are considered typical, and can be made to account for those losses that occur in the blade system. Different manufacturers of steam turbines may neglect to account for some of these but introduce others.

Loss calculation methods are simply a loss accounting system. The sum total of losses, regardless of the number of stages the system contains, should be of the same order. The various methods should result in almost identical losses determined for any stage from a knowledge of its geometry and the local steam conditions.

ANALYSIS OF THE MAGNITUDE OF LOSSES

Having considered the factors that cause or contribute to losses in the steam path, it is of interest to place these in perspective by considering their magnitude both in terms of absolute values and values relative to each other.

Table 3–2 shows the calculated losses in BTU/lb as typical for an impulse type unit rated at 500,000 kW and designed for steam conditions of 2415 psia/1000/1000°F and an exhaust pressure of 2.5 in. Hga. These total losses are divided into three categories. The first contains those losses associated with the stationary blade row, the second those associated with the rotating blades, and the third group losses that can generally be called stage losses. This third group includes the leakage losses around both the stationary and rotating blade elements, disc friction caused by the interaction between the stationary and rotating elements, and the moisture loss determined from the mean stage moisture. These losses are shown diagrammatically for each category in Figure 3–58. These losses from Table 3–2 are shown as three separate groups, the stationary rows *S*, the rotating rows *R*, and general stage losses associated with the stage interaction between the stationary and rotating rows. Also shown are the leaving losses from each section.

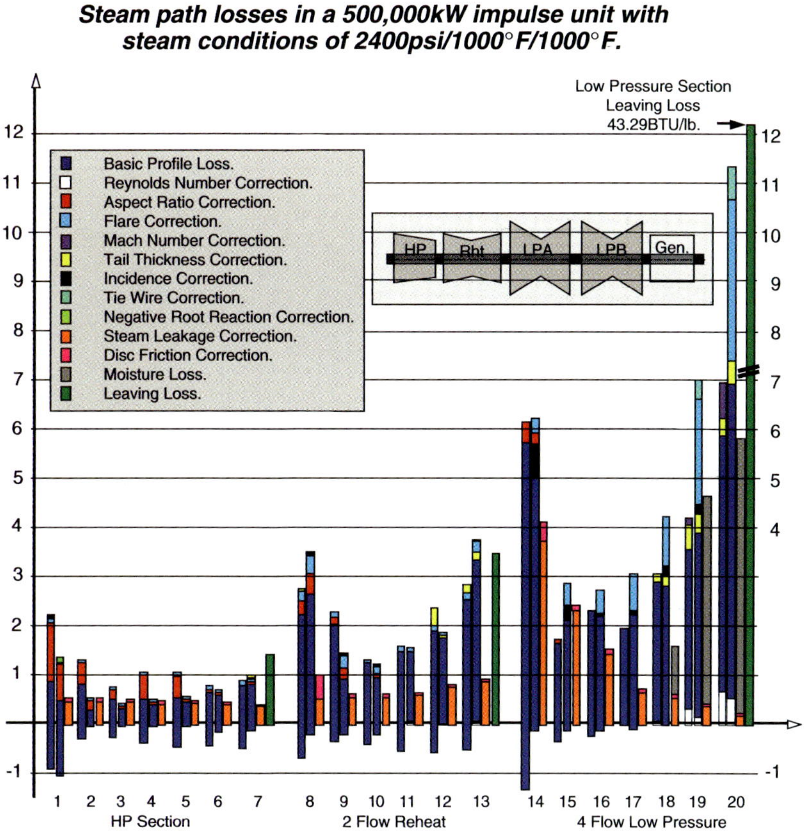

Fig. 3–58 Steam Path Losses by Stage

TURBINE SECTION.			**High Pressure**							**Reheat**		
Stage Number:	1	2	3	4	5	6	7	8	9	10	11	12
Stationary Row.												
Basic.	1.81	1.14	0.74	0.79	0.98	1.14	1.28	2.90	2.35	1.67	2.05	2.46
Reynolds Number.	-0.88	-0.33	-0.28	-0.33	-0.40	-0.42	-0.47	-0.70	-0.37	-0.41	-0.51	-0.56
Mach Number.												
Tail Thickness/Wake.												0.14
Aspect Ratio.	1.21	0.37	0.19	0.49	0.42	0.07		0.33	0.16			
Flare.	0.14	0.05	0.05	0.02	0.07	0.07	0.09	0.19	0.09	0.05	0.07	0.14
Rotating Row.												
Basic.	1.53	0.35	0.33	0.42	0.51	0.70	0.91	2.81	1.14	1.17	1.40	1.74
Reynolds Number.	-1.19	0.00	-0.02	-0.05	-0.05	-0.09	-0.14	-0.20	-0.20	-0.20	0.14	0.09
Mach Number.												
Tail Thickness/Wake							0.09					0.05
Incidence.	0.05					0.02	0.07	0.14	0.05	0.02		
Aspect Ratio.	0.49	0.19	0.09	0.05	0.05	0.05	0.02	0.40	0.26	0.14		
Flare.	0.05	0.02	0.02	0.02	0.02	0.02	0.07	0.28	0.23	0.16	0.05	0.02
Tie Wire.												
Root Negative Reaction.	0.14							0.02				
Stage Losses.												
Leakage.	0.56	0.49	0.49	0.47	0.44	0.44	0.42	0.56	0.60	0.63	0.70	0.79
Disc Friction.	0.14	0.12	0.12	0.09	0.07	0.05	0.05	5.00	0.05	0.05	0.05	0.04
Moisture.												
Leaving Loss.							1.42					
Stage Total	4.05	2.40	1.73	1.97	2.11	2.05	3.81	11.73	4.36	3.28	3.95	4.91
					HP Total		18.12				**Rht. Total**	

Table 3–2 An Analysis of the Steam Path Losses in a 500,000 kW Impulse Design Unit

TURBINE SECTION.				**Low Pressure**					**Total**
Stage Number:	13	14	15	16	17	18	19	20	
Stationary Row.									
Basic.	3.07	7.07	2.00	2.51	2.47	2.79	3.26	5.21	47.69
Reynolds Number.	-0.49	-1.28	-0.14	-0.23	0.00	0.14	0.35	0.72	-6.59
Mach Number.							0.16	0.74	0.90
Tail Thickness/Wake.	0.37					0.19	0.47	0.33	1.50
Aspect Ratio.		0.35	0.05						3.64
Flare.	0.16		0.05	0.33	0.40	0.49	1.00	1.91	5.37
Rotating Row.									
Basic.	3.26	5.16	2.28	2.37	2.34	2.81	3.74	6.37	41.34
Reynolds Number.	0.14	-0.20	-0.19	-0.20	0.14	0.00	0.20	0.56	-1.46
Mach Number.							0.23	1.16	1.39
Tail Thickness/Wake	0.19					0.19	0.37	0.47	1.36
Incidence.	0.05	0.72	0.26	0.05	0.09	0.20	0.19		1.91
Aspect Ratio.		0.21							1.95
Flare.	0.19	0.30	0.37	0.42	0.74	0.95	2.12	3.26	9.31
Tie Wire.							0.33	0.68	1.01
Root Negative Reaction.		0.09							0.25
Stage Losses.									
Leakage.	0.84	3.74	2.33	1.44	0.72	0.56	0.37	0.19	16.78
Disc Friction.	0.02	0.33	0.09	0.09	0.07	0.05	0.03		6.51
Moisture.						1.01	4.23	5.60	10.84
Leaving Loss.	3.44							43.29	48.15
Stage Total	11.24	16.49	7.10	6.78	6.97	9.38	17.05	70.49	191.85
	39.47					**LP Total**		134.26	

Table 3–2 An Analysis of the Steam Path Losses in a 500,000 kW Impulse Design Unit (Cont'd)

As would be expected, a major portion of the loss in all stages is the basic profile loss. This loss is a function of the profile shape and reflects the form of the steam expansion passage it forms. Throughout the expansion, the value of this basic loss ζpo tends to increase. This is because stage available energy increases with increasing diameter so the loss is applied to a larger available energy. The increase in the absolute value of this basic loss does not reflect a degenerating profile shape. The exception to this is the first stage in each section where geometry requirements are different. The profile loss represents about the same percentage of available energy for each stage.

In the earlier stages, higher Reynolds Numbers are achieved as a consequence of the thermodynamic properties of the steam. At these higher values, a thinner boundary layer is achieved on the profile and inner and outer boundaries. This has a positive influence on stage efficiency and reduces losses. The effect of Reynolds Number is influenced by the surface finish of the component as seen in Figure 3–26 where the loss factor χr varies with the ratio of surface finish to chord ks/C.

Within the steam path, flare at the outer boundaries of the low-pressure section latter stages, necessitated by the rate of increase in volumetric flow, causes relatively large losses. This loss is unavoidable because of the need to keep the axial component of steam velocity within acceptable values. Also in the low-pressure sections, as the expansion passes into the moisture region, the moisture frictional drag and impact loss becomes significant.

The leaving losses are shown for all three sections and are relatively small for the high and intermediate pressure ranges although a portion of these losses are recovered as velocity or kinetic energy entering the first stage of the following section. The leaving loss from the low-pressure section cannot be recovered and therefore represents a total loss.

Of the three pressure ranges, 10.2% of the total losses occur in the high-pressure section, 16.6% in the intermediate (reheat) section, and 71.2% in the low-pressure sections. Discounting the leaving losses, and thereby removing any distortion they may introduce due to the selection of low-pressure section annulus area, the fluid path losses occur 12.8% in the high-pressure, 22.4% in the intermediate (reheat), and 65.8% in the low-pressure sections. (Sanders 1982)

It is recognized that steam path efficiency varies along the radial height of each blade row as steam velocities, thermodynamic properties, and structural characteristics of the blade rows vary in the radial direction. This allows an alternate means of viewing the losses in a turbine section. An alternative plot of the efficiency of a low-pressure section is shown in Figure 3–59. (Hong, C.A. "STELLA: A PC-Based Analysis Programme which predicts Performance and Efficiency of Axial Flow Steam Turbines." [Stress Technologies, Inc., Rochester, N.Y.])

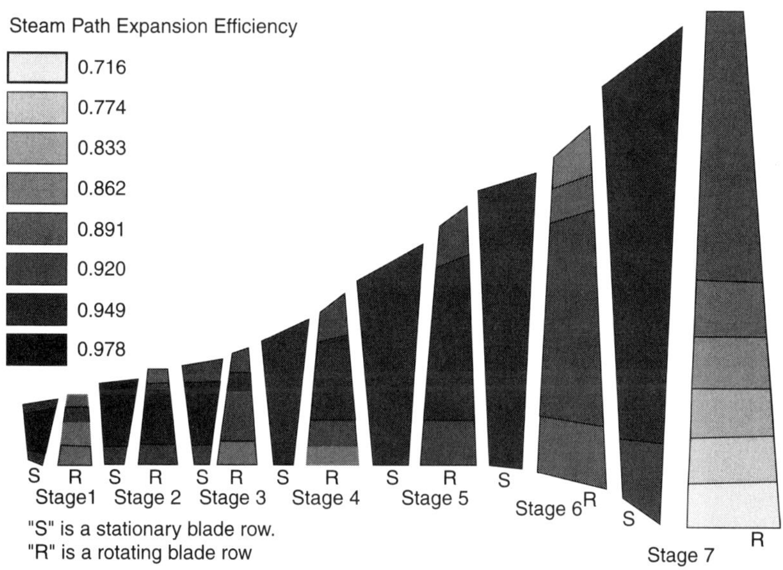

Fig. 3–59 The Efficiency Levels of Stationary and Rotating Blade Rows in a Seven Stage Low Pressure Section

THE COST OF DAMAGE MECHANISMS

A variety of mechanisms and conditions occur within the steam path that cause its efficiency to modify from the original as-delivered quality. While some of these changes can introduce minor temporary improvements in the efficiency of the unit, possibly at the expense of reliability, the vast majority will introduce a reduction. Even those that produce a temporary improvement, such as the thinning of vane discharge edges, will ultimately cause a degradation as more material is removed and the discharge area increase becomes significant.

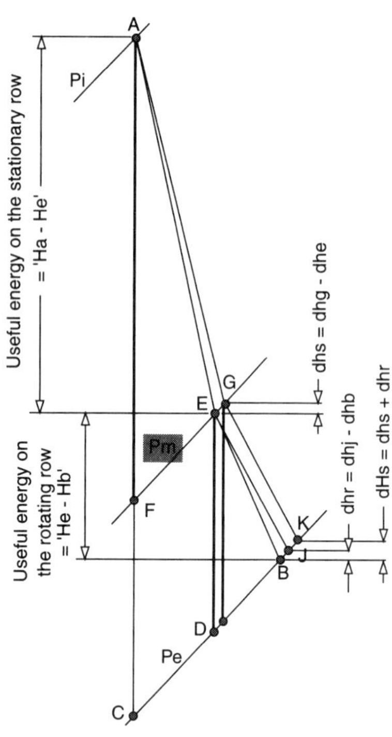

Fig. 3–60 The Effect of Efficiency Deterioration on the Expansion State Line

The cost of damage and deterioration is most conveniently considered in terms of the reduction these phenomena cause in the incremental change in unit output, which is a product of the mass flow and the useful energy. Consider that portion of the Mollier Diagram shown in Figure 3–60. This shows the conditions at inlet to and discharge from a stationary and rotating blade row. That is one stage of a total expansion.

The conditions at inlet to the stationary row are pressure *Pi* and enthalpy *Ha*. After expansion, at design efficiency, the steam has a final condition at discharge from the stationary row of pressure *Pm* and an enthalpy of *He*. The useful energy is then *Ha* - *He*. Any condition that occurs within the steam path and reduces the final steam condition from *E* to *G* represents a reduction in the useful energy of dhs = dhg-dhe. If the steam flow is *M*, then the reduction in stationary row output *dkWs* is:

$$dkWs = \frac{M \, . \, (dhg - dhe)}{3412.14} \tag{3.42}$$

Similarly, the loss of output due to deterioration of the rotating row *dkWr* is found by considering the reduction in useful energy *dhk-dhj.*

$$dkWr = \frac{M \, . \, (dhk - dhj)}{3412.14} \tag{3.43}$$

The total output loss from the stage *dkWt* is the sum of the losses on the stationary and rotating rows, which is equal to the stage mass flow *M* and the sum of the enthalpy losses on the two rows, which is *dhg - dhe* and *dhk - dhj.*

$$\begin{aligned} dkWt = dkWs + dkWr &= \frac{M \, . \, \{(dhg - dhe) + (dhk - dhj)\}}{3412.14} \\ &= \frac{M \, . \, (dhk - dhb)}{3412.14} \end{aligned} \tag{3.44}$$

Once the losses have been reduced to kW output, it is then possible to convert this loss to changes in unit heat rate. By definition:

$$\text{Unit heat rate} = \frac{\text{Thermal Input, BTU/lb.}}{\text{Output in kilowatts.}} \tag{3.45}$$

Therefore, for any change in the unit output there will be a change in the heat rate. The approximate change in heat rate in BTUs per kilowatthour (kWh) can be found from:

$$dHR = HRd \cdot \left[\frac{kWd}{kWd - dkW} \right] \tag{3.46}$$

where

dHR	=	change in heat rate	BTU/kWh
HRd	=	design heat rate	BTU/kWh
dkW	=	loss in kilowatts	kW
kWd	=	design kilowatts	kW

The additional fuel cost on a per annum basis can be found from:

$$\$\,Fc/Annum = \frac{Fc \,.\, \%dHR \,.\, kWr \,.\, LF.\, SHR \,.\, 8760}{10E10} \tag{3.47}$$

where

$Fc/Annum	=	annual fuel cost increment	$/Annum
Fc	=	fuel cost	c/10E6 BTU
%dHR	=	% change in heat rate from 3.46	
kWr	=	unit rated kilowatts	kW
LF	=	unit annual load factor	
SHR	=	station heat rate	BTU/kWh
8760	=	number of hours in a year	hours

Alternatively equation 3.47 can be expressed as:

$$\$Fc/Annum = Ks . Fc . \%dHR \tag{3.48}$$

where

$$Ks = \frac{kWr . LF. SHR . 8760}{10E10}$$

Another method of expressing this additional cost to station operations is to calculate a value of incremental dollars per megawatt hour (\$/MWh) of generated output using the expression:

$$\$ IG = \frac{Fc . \%dHR . SHR}{MW . LF . 8760} = \frac{\$ Fc/Annum}{MW . LF . 8760} \tag{3.49}$$

where

MW = unit rating in MW

Example 3.1

A unit rated at 500,000 kW has a design station heat rate of 9650 BTU/kWh and an annual load factor of 0.83. If the steam path losses have been determined to be 1170 kW, determine the annual additional fuel cost for a range of fuels from 150 to 300 cents per 10E6/million BTU.

Solution:

Applying equation 3.46:

$$dHR = 9650 \times \left[\frac{1170}{500000 - 1170}\right] = 22.63 \text{ BTU/lb.}$$

$$\%dHR = \frac{dHR}{HRd} \times 100, \quad = \frac{22.63}{9650} \times 100 = 0.235\%$$

From equation 3.49:

$$Ks = \frac{500000 \times 0.83 \times 9650 \times 8760}{10E10} = 3508.16$$

Fuel Cost	%dHR	$/Annum	$I/ kW loss
150	0.235	123,662	105.69
200	0.235	164,884	140.93
250	0.235	206,104	176.16
300	0.235	247,325	211.39

The losses in the steam path are numerous, and it is unlikely the output of the unit in its mint condition can be maintained for a significant period after it has been placed in service. In addition, once this pristine condition is lost, it can never be recovered completely. There is a certain amount of refurbishment and restorative work that can be undertaken at a maintenance/inspection outage, and the need for the operator is to determine how much of this work can be justified, remembering that much of the cleaning and upgrading to, or near, the original condition will be retained for only a relatively short period.

The following areas of loss represent those most often noted when a unit is opened and are those on which corrective action can be taken. The extent that the improvement can be maintained is dependent upon the form of deterioration, the corrective actions taken, and to a degree the manner in which the unit is operated after return to service.

Loss due to steam leakage (bypassing of blade elements)

Steam that leaks under stationary or over rotating blades does no work on them. In addition this steam will at most locations reenter the main steam path causing further losses by introducing disturbance to the orderly flow of the main working fluid. These losses are difficult to quantify but their presence and influence are recognized by the design engineer.

Loss due to steam leakage at shaft-end glands

Steam that leaks from the turbine section at shaft-end packing positions does not in general reenter the steam path but is directed to some lower energy location within the working fluid system such as a feed water heater.

There are certain considerations relating to this shaft-end packing that need to be considered. These include:

- The steam leaking through shaft-end packing will bypass lower enthalpy, downstream stages and therefore do no work within the unit.
- The leakage steam is directed to the highest energy position possible so that as much energy as possible can be extracted from it.

If this steam reenters the steam path, then it will probably cause reentry loses in much the same manner as that steam which bypasses the blade system and discussed previously.

Loss due to surface deposits

The deposition on the blades and the inner and outer sidewalls results in roughening that causes losses to increase. These losses cannot be avoided, but by careful control of the plant chemistry, their harmful effects to efficiency can possibly be delayed. However, only small levels of deposits on the expansion surfaces are required to cause high loss ratios.

Determination of losses due to profile damage

Damage mechanisms that cause any form of surface deformation of the blade vanes will introduce losses into the expansion efficiency. Craters such as shown in Figure 3–61 will cause flow separation from the vane surface and turbulent flow. Similarly, deformation of the vane as shown in Figure 3–62 will cause losses by causing disruption to the steam flow.

Fig. 3–61 Minor Impact Craters on the Suction Surface of a Stationary Blade Row of an Impulse Unit

Fig. 3–62 Vane Deformation Due to Solid Particle Impact

Losses due to discharge area deformation

Any action that causes a change in the discharge area of a stationary or rotating blade row will influence the row pressure distribution, modifying the enthalpy drop, and the efficiency. There are various mechanisms that can either increase or decrease this area, and they can be equally destructive of efficiency.

INFLUENCE OF STATE-LINE EFFICIENCY ON HEAT RATE

To determine operating cost changes as a function of time, it is necessary to be able to determine state-line efficiencies and flow quantities with some level of confidence that the values are sufficiently accurate to be used for making maintenance and repair decisions.

The expansion efficiency of a steam path, if the total expansion is in the superheat region, can be determined with considerable accuracy by means of an enthalpy drop test using high quality instrumentation to measure steam pressure and temperature. (Cotton 1993) Such a test will establish steam conditions at entry to and exit from a section. A more difficult parameter to establish is the expansion efficiency in a low-pressure section, and only by a detailed audit can this be determined. Even after such an audit, the results leave some degree of uncertainty as to the exact value, and for an audit to be conducted the unit needs to be open for a detailed examination. However, the results of an audit can be used with some level of confidence. But in an operating unit, it is generally the experience of station personnel that is relied upon to determine, or estimate, the low-pressure section condition, normally from past experience and observed changes in operating parameters.

The flow quantities to each section and the quantities being extracted are other parameters that are difficult to establish with any degree of accuracy, but if calculated values taken from the heat balance are used, an estimate of the effect of deterioration on unit or station heat rate can be determined. Then, by applying equation 3.47 or 3.48, the incremental fuel costs can be established.

Using the data discussed here and shown in Figure 3–63, it is possible to establish heat utilization factors for each section using the steam quantities and thermal conditions around them. If the loss factor in the high-pressure, reheat, and low-pressure sections are designated *Fh, Fr,* and *Fl,* these can be determined from:

$$Fh = \frac{[Mhi + Mho] \times [Hhi - Hho]}{2.\Sigma\ M.H}$$

$$Fr = \frac{[Mri + Mro] \times [Hri - Hro]}{2.\Sigma\ M.H}$$

$$Fl = \frac{[Mli + Mlo] \times [Hli - Hlo]}{2.\Sigma\ M.H} \qquad (3.50)$$

where

- M is the steam flow in lb/hr
- H is the enthalpy in BTU/lb
- i refers to inlet conditions
- o refers to outlet conditions
- h refers to the high-pressure section
- r refers to the intermediate pressure section
- l refers to the low-pressure section

In equation 3.50, the value of the term $\Sigma M.H$ can be determined from:

$$\Sigma MH = Mh.\left[Hhi - Hho\right] + Mr.\left[Hri - Hro\right] + Ml.\left[Hli - Hlo\right] \tag{3.51}$$

In equation 3.51, the terms *Mh, Mr,* and *Ml* are the mean flows through the high-pressure, reheat, and low-pressure sections. They are determined from:

$$= 1/2\left[Mi + Mo\right] \tag{3.52}$$

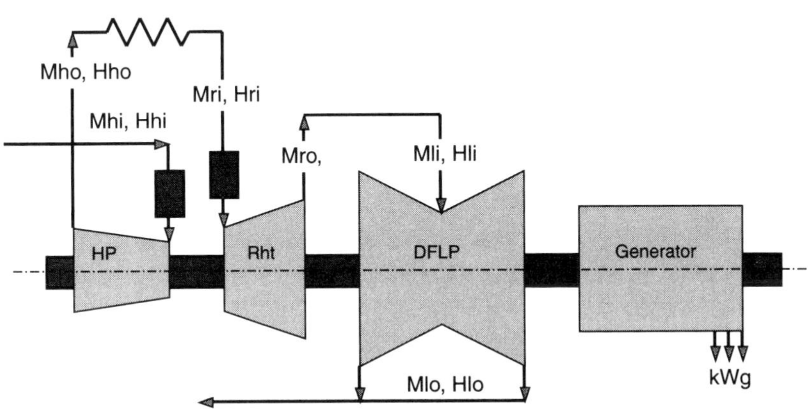

Fig. 3–63 The Nomenclature Around a Reheat Unit Showing the Steam Flows and Enthalpy

Where sections have steam extracted for regenerative feed heating or any other purpose, it is more accurate to use the enthalpy for individual blade groups and the flow through them. However, unless there is absolute confidence in the values determined for flow *M* and the state-line efficiency, there is little advantage to making this minor correction. Also, in multi-flow low-pressure sections with non-symmetric extractions, the problem becomes even more complex.

Example 3.2

As an example, consider a 500,000 kW unit in which an enthalpy drop test indicates there is a deterioration the in high-pressure section of 1.750% in state-line efficiency and a further deterioration of 2.04% in the reheat section. Assume also there are sufficient indications from previous inspections that a 2.00% loss is estimated for the low-pressure section. Using this data, the total heat rate degradation can be found.

This unit also has a station heat rate (SHR) of 9845 BTU/kWh and consumes fuel that costs 275cents/10E6 BTUs. The unit load factor is 0.763. Determine the annual additional fuel costs if these values are unchanged for the next 12 months.

Using the steam-cycle parameters shown in Figure 3–64 and applying equation 3.50 to this cycle, the deterioration factors *F* can be determined.

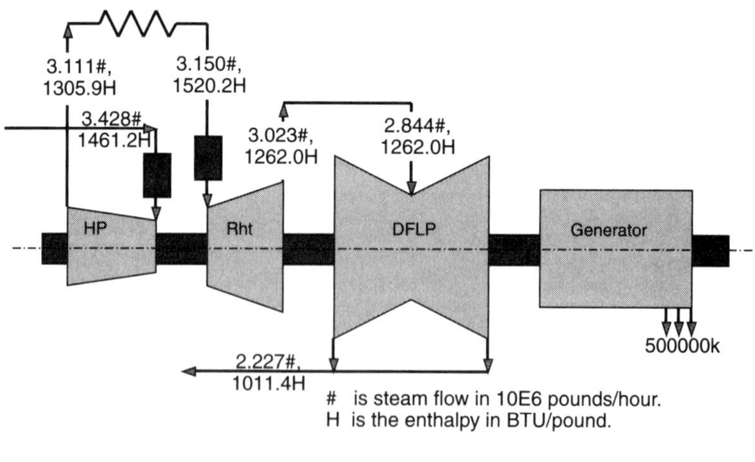

Fig. 3–64 Steam Conditions Around Unit of Figure 3–63

$$M_h[H_{hi} - H_{ho}] = 1/2[3.428 + 3.111] \times [1461.2 - 1305.9] = 507.75 \times 10E6 \text{ BTUs}$$

$$M_r[H_{ri} - H_{ro}] = 1/2[3.150 + 3.023] \times [1520.2 - 1262.0] = 796.93 \times 10E6 \text{ BTUs}$$

$$M_l[H_{li} - H_{lo}] = 1/2[2.844 + 2.227] \times [1262.0 - 1011.4] = 635.40 \times 10E6 \text{ BTUs}$$

$$\Sigma MH = (507.75 + 796.93 + 635.40) \times 10E6 = 1940.08 \times 10E6 \text{ BTUs}$$

Therefore, the heat degradation factors *F* can be found from:

Fh = 507.75/1940.08 = 0.2617
Fr = 796.93/1940.08 = 0.4108
Fl = 635.40/1940.08 = 0.3275

Applying these degradation factors to the measured and gauged efficiency deterioration allows the following heat rate deterioration to be determined:

High-pressure section.	(1 - 0.0175) x 0.2617	=	0.2571
Intermediate-pressure section.	(1 - 0.0204) x 0.4108	=	0.4024
Low-pressure section.	(1 - 0.0200) x 0.3275	=	0.3210
Total change (reduction) in unit heat rate		=	0.9805%

Therefore, the heat rate will increase to *1/0.9805* or 1.01989 giving a 1.989% increase.

This change in cycle heat rate will require extra fuel be consumed to develop the same output, and the additional fuel cost (or fuel cost penalty) per annum can be calculated using equation 3.47:

$$\$\,Fc/Annum = \frac{Fc\,.\,\%dHR\,.\,kWr\,.\,LF.\,SHR\,.\,8760}{10E10}$$

$$= \frac{275 \times 1.989 \times 500000 \times 0.763 \times 9845 \times 8760}{10E10}$$

$$= \$1{,}799{,}624/Year,\ or\ \$4{,}391 \cong \$\,4{,}900\,/Day.$$

An alternate method of expressing this additional cost to station operations is to calculate a value of incremental \$/MWh from equation 3.49:

$$\$\,IG = \frac{\$\,Fc/Annum}{MW\,.\,LF\,.\,8760}$$

$$= \frac{1{,}799{,}621.1}{500 \times 0.763 \times 8760} = 5.385\ \$/MW - Hr.$$

REFERENCES

Abbot I.H. and A.E. von Doenhoff. *Theory of Wing Sections.* New York: Dover Publications, Inc., 1958.

Ainley, D.G. and G.C.R. Mathieson. "An Examination of the Flow and Pressure Losses in Blade Rows of Axial Flow Turbines," Aeronautical Research Council, R & M 2891, 1955.

———. "A Method of Performance Estimation for Axial Flow Turbines," Aeronautical Research Council, R & M 2974, 1957.

Baily, F.G., K.C. Cotton and R.C. Spencer. "Predicting the Performance of Large Steam Turbine-Generators Operating with Saturated and Low Superheat Steam Conditions." Presented at the 29th Annual Meeting of the American Power Conference, April 1967, Chicago, IL.

Chen, S. "A Loss Model for the Transonic Flow Low Pressure Steam Turbine Blades."

Cotton, K.C. *Evaluating and Improving Steam Turbine Performance.* New York: Published by Cotton Fact Inc. Rexford, 1993.

Craig, H.R.M. and H.J.A. Cox. "Performance Estimation of Axial Flow Turbines," Proceedings of the Institute of Mechanical Engineers. 1970-71.

Dewey R. and Rieger N. F. "Blade Failure," EPRI Survey.

Forster V.T. "Performance Loss of Modern Steam Turbine Plant due to Surface Roughness." Proceedings, Institute of Mechanical. Engineers, Vol. 181, Part 1, Number 17. 1966-67.

Hong, C.A. "STELLA: A PC-Based Analysis Programme which predicts Performance and Efficiency of Axial Flow Steam Turbines." Stress Technologies, Inc., Rochester, N.Y.

Horlock, J. H. *Axial Flow Turbines.* London: Butterworths, 1966.

Kearton, W.J. *Steam Turbine Theory and Practice, Seventh Edition.* London: Sir Isaac Pitman & Sons Ltd., 1958.

Markov, M.H. "Calculation of the Aerodynamic Characteristics of Turbine Blading," translated by Associated Technical Press, 1958.

Martelli, F. and A. Boretti. "Development of an Experimental Correlation for Transonic Turbine Flow." ASME Paper 86-GT-108, 1986.

Martin, H.M. "Steam Turbines." *The Engineer,* London, 1913, p. 1610.

Milne-Thomson L.M. *Theoretical Aerodynamics, 3rd Edition,* London: McMillan & Co. Ltd., 1958.

Prandtl,L. and H. Schlichting. *Das Widerstandsgesetz rauher Platten.* Werft, Reed. Hafen, 1934, 1.

Ribrary. "The Frictional Losses of Steam Turbine Discs." *The Brown Boveri Review,* 1934.

Salisbury, K. J. *Steam Turbines and Their Cycles.* Huntington, N.Y.: Robert E. Kriger Publishing Company, 1974.

Sanders, W.P. "Turbine Steam Path Efficiency," Canadian Electrical Association, Regina, Saskatchewan, October 1982.

Smith, A. and H.R. Gibson. "The Measurement of Turbine Blade Loss in Cascade," *Parsons Journal,* Christmas, 1964.

Spencer, R.C. Cotton K.C., and C.N. Cannon. "A ,Method for Predicting the Performance of Steam Turbine Generators, 165,000 kW and Larger," ASME Paper 62-WA-209, Annual Winter Meeting, New York, NY, 1962.

Stodola, *Dampf und Gas Turbinen, 6th Edition.* 1924.

Traupel, W. *Thermische Turbomashinen,* Berlin, Springer Verlag, 1966.

Chapter 4

Performance Testing

INTRODUCTION

Testing undertaken on a steam power cycle can have several objectives.

- When the unit is new or has just been installed, tests are conducted to establish the efficiency or heat rate to ensure guarantee levels have been reached. These tests may be undertaken at various loads conditions depending upon the guarantee at different steam conditions.
- As the unit ages and deterioration occurs, the unit may be tested to determine both the extent of deterioration and to identify areas where this deterioration has occurred so that corrective action can be planned.

These performance tests should be run in strict accordance with the test code to minimize test errors to the greatest extent possible.

A major consideration with inefficient operation is identifying and quantifying the source of losses. Output tests can help establish the magnitude of such losses, but the determination of their source and individual magnitudes is often difficult and a detailed steam path audit is required to quantify and assist plant staff in maintenance decisions. Once the losses have been identified and their magnitude quantified, a financial penalty can be assigned to them. It then becomes easier to make replacement/repair/accept-as-is decisions based on fuel costs, predicted load factors, and other characteristics of the unit operation. There are also considerations related to the availability of replacement parts, special tools, and skills that may be necessary to implement the corrective actions. Each of these must be fully evaluated by operators in determining the most cost-effective approach for correction and return to service, plus the need and advisability of taking corrective action.

THE PURPOSE OF PERFORMANCE TESTING

The level of efficiency at which a steam turbine-generator operates is of considerable importance to the owner as this reflects the fuel costs he must pay for the energy generated. Efficiency is important in the new unit and then the rate over which this varies or deteriorates during the unit's operating life.

At the completion of erection or installation and when the unit has been commissioned, a performance acceptance test is often run. The primary objective of this test is to measure the energy conversion efficiency level and compare this test or measured value to that guaranteed by the supplier. The comparison is made for both heat rate and capacity. A performance test can also have the secondary objective of providing meaningful information to the manufacturer regarding the performance level of a particular section of the unit where design innovation has occurred. Such tests can also be designed to provide information about a particular component within a section.

However, this is usually achieved only by the use of instrumentation beyond what is normally required for an acceptance test.

When a performance test is run, it is in the interest of the owner and manufacturer to take all precautions necessary to ensure measurement inaccuracies are minimized and a meaningful accurate result is obtained. This is because this data, in addition to establishing the heat rate or efficiency of the unit and cycle, will provide the owner with a reference point against which the future performance can be gauged. An accurate test will also allow the performance of individual sections of the unit to be defined and monitored to provide an indication when it has deteriorated to an unacceptable degree and when corrective action should be taken. Such comparison of later test data with the original test results can also provide an indication of mechanical damage or deterioration. For designers, good tests apprise them of the performance level of the unit and enable them to calibrate or modify the designs. In the case of prototype units, it will help establish their performance level and possibly indicate or identify areas where unit performance could be improved.

An accurate, well-conducted test can indicate or identify areas in the turbine or cycle where performance, although apparently normal, is below the predicted or guaranteed level and where remedial action can possibly rectify the situation or prevent further deterioration. Test information can also provide information that can be used to assist in reducing generating costs. These considerations alone may be more than sufficient justification for the expenditure involved in performing an acceptance test. This is particularly so in the case of a prototype design with components for which there may be only limited field data available or where predicted performance levels are based on experimental work that needs to be verified.

Often, when using a prototype design or when a manufacturer is utilizing new components or a new arrangement of existing components, the manufacturer is most anxious to undertake a detailed test to allow him to calibrate his experimental results to actual field data. Such information provides the manufacturer with a database, allowing calibration of those design adjustments that may be required and also to more accurately predict and guarantee results on subsequent units. Often when new components are used, the user may participate in the test and even provide later data to the manufacturer.

One very important aspect of performance testing is that such tests enable a manufacturer to demonstrate that the unit will meet the guaranteed levels of performance on which the unit was sold. These results will also enable the purchasers to accumulate information regarding the probabilities of various manufacturers achieving guaranteed performance. This information will enable the owner to assess more realistically the heat rates being offered on future units and allow a better assessment in evaluations for subsequent capacity purchases. Shown as Figure 4–1 is a shotgun plot of test results without any adjustment for testing margin or other factors for one manufacturer. (Neustaedter and Sanders 1969) This plot makes a comparison of the guaranteed to actual test-determined values of heat rate to the guaranteed level. The tests were conducted according to the ASME PTC-6 Power Test Code for Steam Turbines.

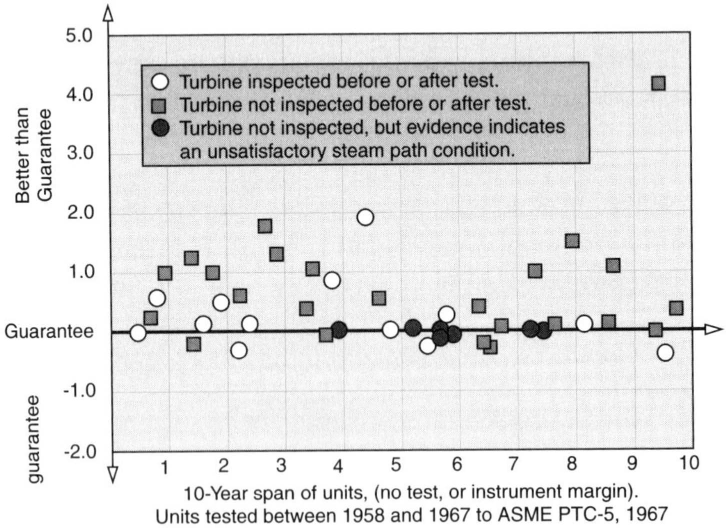

Fig. 4–1 Shot Gun Plot of Performance Test Results for a 10-Year Period

For a performance test to accurately establish the efficiency of the steam turbine-generator, three parameters must be accurately measured.

1. **Steam quantity flowing in the cycle and turbine unit.** The quantity of steam flowing varies at different positions within the cycle and turbine due to designed extractions and flow splitting into parallel paths. Even knowing the quantity of steam within a section does not provide absolute knowledge of the quantity generating power because there are internal leakages within the stages that modify the actual flow through the blades. However, these leakage flows represent losses, and the calculated efficiency does not require an absolute knowledge of the quantities leaking.

2. **Thermal energy available and supplied from the boiler.** This energy is established through measurements of the steam conditions in terms of pressure and temperature at various locations within both the turbine and cycle, most particularly at inlet and discharge from the turbine sections. The total energy at any stage position within the turbine unit is the sum of two distinct forms of energy. These are the thermal potential energy and the kinetic energy of the steam as it passes through the blade rows. This kinetic energy is relatively small or practically zero in most measurable positions within the steam path. However, at the exhaust position in the low-pressure section this is of sufficient magnitude it is important.

3. **Power generated within the unit.** The expansion of steam produces power in the steam path that is transmitted through the rotor to drive the generator. There are various mechanical losses within the turbine-generator unit, and therefore the power developed in the generator is equal to the power developed in the turbines minus the power they absorb in overcoming friction, windage, and other mechanical losses. The generator power is further reduced by the electrical

power required to overcome internal losses within the generator such as heating and hysteresis losses and also the power required to excite the generator.

Therefore, a major requirement of any performance test is to accurately measure these parameters, both the primary values and the secondary effects which act to influence these primary values. Thermal power is measured by pressure and temperature, flow by metering, and power by measurement of electrical power output.

After testing and measurement of the performance values, it is necessary to make certain corrections for the steam-cycle conditions and the performance of equipment that forms part of the thermal power conversion cycle but is not part of the steam turbine generator supply. This equipment can comprise pumps, motors, heaters, or other heat transfer equipment. To make this correction will require either careful determination of their individual performance as part of the test, and then correction to reflect their true performance or their careful calibration prior to its use. Steam cycle parameters are measured with calibrated instrumentation and then the test results corrected to reflect the true readings.

In addition to determining the efficiency levels of the unit tested, performance tests can be extended to prove the units are capable of operating under emergency or abnormal conditions. In this respect, the characteristics of the unit are studied to determine their suitability for the system and to make any final adjustments to the control mechanisms and operating systems.

After an analysis has been made of the test results and efficiency levels determined—possibly at various loads—these are compared to the guaranteed level of performance. There are certain courses of action that may result depending upon the contract under which the unit was purchased. These include:

- a monetary settlement between the supplier and owner if the contract contains a bonus/penalty clause
- if the unit has failed to achieve the guaranteed level of performance, then the manufacturer has the opportunity to modify details to bring it to guaranteed performance level

For these reasons, it is necessary to reduce test inaccuracies to the greatest extent possible and eliminate the level of uncertainties that exist in the calculated values of unit and cycle efficiency. For this reason the instrumentation chosen, its location, and connection within the system are critical. Details of the instrumentation requirements will be discussed later in this chapter.

The data from testing represents an as-tested condition. These conditions may not be the same as the design-specified requirements, so they will tend to affect the accuracy of the test. Therefore, it is necessary at completion of the test to correct possible cycle differences and differences in the various portions of the equipment back to an as-designed condition. The corrections normally relate to the various components of the cycle as well as the test steam conditions. The correction factors should be agreed between the parties before the test is conducted.

TEST COSTS AND ECONOMICS

The cost of conducting a performance test to code standard is relatively high. There are costs associated with the need to make the unit available for the connection and removal of instrumentation and the cost of providing personnel to conduct the tests. There may even be costs associated with disruption to the system, and this is before the unit has been in service for a significant period on a system that normally requires the power the units will develop. However, the efficiency with which energy is converted in the unit must be considered. The newer units on a system are normally the most cost effective, having taken advantage of improved energy conversion technology when compared to the older units which are available and having been selected to burn the most cost effective fuel. Therefore, performance tests are a major cost item. However, if the code test results are used to identify deficiencies in the unit, to identify upgrading needs, or provide an accurate indication of deterioration or mechanical difficulties, then such costs can be easily justified.

Fuel costs are a major expenditure within the total cost of operating a power generating facility. Therefore, the determination and monitoring of unit efficiency and its rate of deterioration should be a major consideration in optimizing plant operating costs. For this reason, if a performance test is not conducted when the unit is new, then there is a less substantial base on which to gauge deterioration and total fuel costs. That is, the initial test results act as a base or reference point against which to judge the subsequent operation of the unit.

In operating a plant effectively, it is necessary to establish the cost of efficiency deterioration and to provide a means of establishing additional fuel costs that occur as a result of such deterioration. Consider a unit rated at 600,000 kW, with a station heat rate of 9500 BTU/kWh. If this unit operates for one year with a load factor of 75%—that is 0.75 x 8760 = 6570 operating hours—then the annual fuel cost penalty for a 0.10% station heat rate deficiency would be:

$$\text{Annual Incremental Fuel Cost '\$'} = \frac{600{,}000 \times 6570 \times 0.10 \times 9500 \times FC}{10\text{ e }10}$$

$$= 374.5 \times FC \text{ \$/Annum/0.10\% deterioration}$$

This example gives fuel costs (FC) in cents per million BTUs.

If the plant is burning coal at 150c/million BTUs, then the annual fuel cost penalty associated with a 1.0% deterioration in heat rate is

$$\frac{1.0}{0.1} \times 374.5 \times 150 = \$561{,}750/1.0\% \text{ deterioration.}$$

If the fuel is oil at 300c/million BTUs, then the annual fuel cost penalty associated with a 0.1% heat rate deterioration is

$$\frac{1.0}{0.1} \times 374.5 \times 300 = \$1{,}123{,}500/1.0\% \text{ deterioration.}$$

Because the station heat rate is 9500 BTU/kWh, a 1.0% deterioration represents a 95 BTU/kWh loss and an additional requirement for fuel consumption for an equivalent output. This is a large amount and represents for the fossil fuels considered above a loss of $561,750/annum for coal and $1,123,500/annum for oil.

From this brief analysis, it can be seen the identification of an efficiency shortfall in a new or retested unit can have a considerable impact on maintenance decisions that must be made when a unit is to be removed from service. It will also influence the level of replacement parts purchased and installed. Such test information can also influence the decisions made as to when a unit should be removed from service prior to its planned maintenance period for restoration.

CLASSES OF PERFORMANCE TEST

There are various classes and/or levels of performance test that can be conducted. The following classification is provided for convenience and is not intended to be exhaustive. Rather, these levels are intended to represent those options available to owners and manufacturers and to indicate the level of reliance that can be placed on the results achieved. These possible levels of test will be described in the order of the accuracy, and therefore the level of credibility that can be placed on their results, and in general, the magnitude of the expense associated with conducting them.

A performance test is normally intended to provide an accurate guide to the performance level of the unit but can be conducted for reasons other than arriving at a simple comparison to the guaranteed level. Dependent upon the reason for conducting the test, the costs associated with running it can be carried by one party or borne by the purchaser and supplier in part.

A test requested by the manufacturer

A performance test is often requested by the manufacturer to obtain design-specific information sufficient to calibrate the design of components or their arrangement or to verify in-house testing and calculations. Such tests are generally of the entire unit—otherwise it could not truly be classified as a performance test. In certain circumstances, it could be a detailed test of one particular portion, or even components of the unit, in which the section or a particular portion of the unit is isolated and instrumented to allow its performance to be identified in detail.

Because such a test is of considerable interest to the manufacturer, who is in effect using the owner's equipment and facilities to accumulate field data, the cost of such a test is normally paid for largely by the manufacturer. It can often happen that the extent of the tests will be extended beyond the requirements for establishing performance levels, and this could extend the test period. These are matters that the manufacturer will explain to and negotiate with the owner.

In some applications of a new component, a manufacturer will incorporate new design features into a unit that is to be tested on the understanding that a detailed test will be conducted. Similarly, manufacturers will delay the installation of a design feature into a unit until agreement can be reached with a purchaser prepared to sanction and participate in such a test.

It is not uncommon for a manufacturer to incorporate a detailed component test into a suitable unit on which the owner wishes to conduct a test to ensure compliance to the standards of a code. In this manner, the manufacturer and owner will obtain detailed information of both the complete unit and the component being studied. A question often encountered in such a test is the proprietary nature of the specific test data acquired concerning the component. It is normal for the manufacturer to reach complete agreement on this matter before any such test is undertaken. There are instances where design modifications may be made to one portion of a unit or to a complete unit with a sister unit in the same plant so direct comparisons can be made.

Performance test conducted to a specific test code for guarantee purposes

This class of test is normally undertaken by the manufacturer and usually at the request and expense of the purchaser. This is done to determine the performance of the unit for guarantee and possibly monitoring purposes. Such tests are normally conducted in strict agreement to a specific code, and there is a need for meetings between the parties to the test so matters relating to its satisfactory undertaking and individual responsibilities are agreed upon well in advance of the unit being delivered.

To conduct such a test, there are special provisions to be made in piping and other equipment of the cycle to ensure the greatest degree of accuracy possible is achieved. It is necessary that many contentious issues relating to a test be resolved early in the contract, as their impact on test schedules and the manner in which they are conducted can affect the total installation. Such tests are normally termed *code tests.*

Such a code test cannot be designed into a system on short notice. It is normally necessary for the purchaser to define the requirements of such a test in the purchase specification. This is because it would not be possible to install the quantity, type, and accuracy of instrumentation required on short notice. It would also be necessary to have access to portions of the cycle equipment so special provision can be made to install the necessary instrumentation. Such access also requires that many special connection points be made available in areas that are not accessible after fabrication is complete.

In a test conducted for guarantee purposes, it is necessary for the test points (load levels) to be agreed upon and any formula required to determine a mean or weighted heat rate must be defined. Penalties associated with deficient heat rates or output are normally defined in the purchase specification and a cost assigned to such deficiency.

Performance test to code standards conducted by the owner

This test is essentially the same as that just discussed. Here the test requirements are just as stringent. The details of test duration, steam parameter fluctuation, and test instruments are identical. However, in this case the test is performed by the owner.

If this test is intended as the basis of a guarantee with the possible consequence of penalties, it is normal for the manufacturer to have a representative present during the test. This representative will normally be an engineer who conducts such tests for the manufacturer. The manufacturer's representative will also check test instrumentation, calibration, and procedures to ensure they conform with the code against which the unit is guaranteed. Because of guarantee implications, it is considered important by the manufacturer to ensure the test accurately reflects the quality of the turbine-generator.

For such a test, it is still normal for the manufacturer to request test meetings. These meetings will ensure that details of the test are defined and that the test will be conducted in such a manner the manufacturer can agree with test procedures and the results obtained.

A test not conducted to a specific code

A test under this category does not follow any particular code in detail, and normally the manufacturer will not participate in or recognize the results as forming the basis for any guarantee unless such a test has been given prior recognition by the contract. However, this is not a condition normally accepted by manufacturers, as the implications can be severe.

The most often stated reason for undertaking such a test is to reduce the cost of conducting it. Also to a certain degree, the reduced time requirements to undertake the test because such a test will not normally be as rigorous as a code test.

A reason often stated for such a test is that it provides a bench mark against which future deterioration can be monitored. However, the results of such a test can have a level of uncertainty outside the level of unit degradation that can be expected in normal operation. For this reason, it is difficult to state with any degree of certainty that the unit condition can be judged from the comparison of results of subsequent tests, even if they are performed with the same instrumentation, and all other conditions are made as nearly compatible with the initial test as possible.

The technical value of such a test is questionable, and although the results may be obtained at less cost, it is normally difficult to justify the use of these values as a basis of unit calibration. Also the repeatability of the results is very questionable.

If station instrumentation is used, there is a question of the overall accuracy of the test, and it is doubtful that these results will allow meaningful unit monitoring to be undertaken because station instrument calibration will vary with time, and it is not known to what level of recording accuracy the original test was run.

This form of test should not be considered suitable for any detailed technical evaluation of the unit or cycle and will not be discussed further.

TEST OBJECTIVES AND REQUIREMENTS

The primary objective of a performance test is determining the efficiency or heat rate of the thermal cycle and specifically of the steam turbine-generator with a degree of accuracy such that the results represent a true determination of these parameters. To do this, the test must measure and quantify those quantities of the working fluid that are required to allow the heat rate to be calculated. To establish what these quantities are, consider the equations of heat rate for two typical cycles. Shown in Figure 4–2 is a simple cycle employing a non-reheat turbine-generator with five stages of regenerative feed water heating and having a boiler feed pump (BFP) between the deaerator and top heater. The positions and requirements of these measurements are shown.

$$\text{Net Heat Rate} = \frac{Qt\,.\,(Ht - hfw)}{kWg - kWp} \tag{4.1}$$

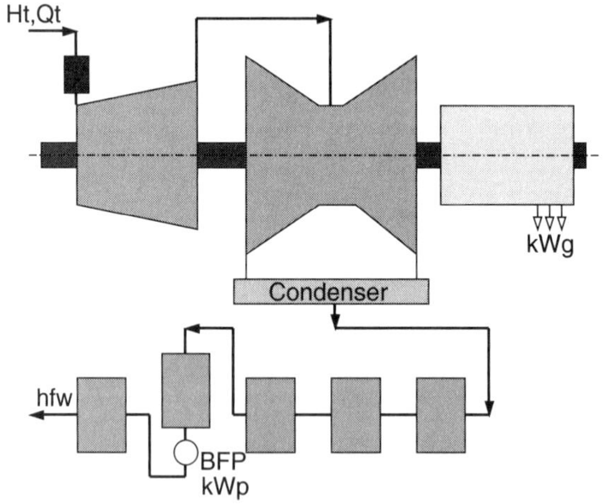

Fig. 4–2 Required Measurements of Steam Properties and Power for the Test of a Simple Non-Reheat Unit

Shown as Figure 4–3 is a more complex cycle employing two stages of reheat with steam extracted from reheat section one and reheat section two to have their enthalpy level raised. This cycle has eight stages of regenerative feed water heating comprising four low-pressure heaters, a deaerator, and three high-pressure heaters.

$$\frac{\text{Net Heat Rate}}{\text{(Motor Pump)}} = \frac{Qt\ .\ (Ht - hfw) + Qr1\ .\ (Hhr1 - Hcr1) + Qr2\ .\ (Hhr2 - Hcr2)}{kWg - kWp} \tag{4.2}$$

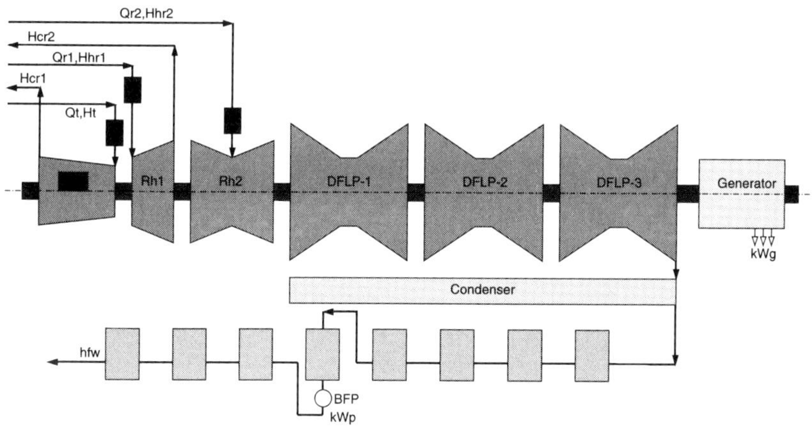

Fig. 4–3 Measured Test Requirements for a Double Reheat Unit

These two figures show the quantities required to be established to allow the cycle heat rate to be established. The enthalpy is determined in terms of the pressure and temperature at any location. These measured values are considered primary parameters and the test objective is to establish these parameters with a high degree of accuracy to eliminate test inaccuracies as much as possible.

THE ENTHALPY DROP TEST (EDT)

An integral part of many performance evaluations that is normally conducted as a precursor to a full performance test, is the EDT. Such tests can be conducted on any section of a unit that is operating entirely on superheated steam throughout the entire expansion range. These tests are used to establish the internal state-line or expansion efficiency by making accurate measurements of the enthalpy at inlet to and discharge from the section. The difference in these two values of enthalpy provides the useful heat energy within the section. This useful energy is then compared to the available or isentropic heat energy between these same two pressures. The ratio of these two energies, provides the state-line efficiency, which was discussed in more detail in chapter 1.

This EDT is relatively inexpensive to run—it can be conducted within a short time and without the need to remove the unit from service. This test is an excellent means of monitoring the performance of individual sections, and many owners use such EDTs to monitor their units to establish the extent to which they have suffered deterioration. This efficiency data can then be used to assist in planning maintenance outages.

Conducting an EDT

The EDT is a relatively short duration test, and normally instrumentation of adequate accuracy can be set up, the test conducted, and the instrumentation removed in a single day. This can be done without interfering with the operation of the plant, except that the steam conditions and load must be held constant within test specifications for the duration of the test (normally about two hours).

The requirements of such a test are:

- Test Grade instruments capable of measuring pressure and temperature must be used. It is normal to employ duplicate instrumentation at both inlet and outlet positions on each section to be tested and to use the same connections as used for a performance test. That is, the test instrument location and placement must be the same as required for a full code performance test.
- A test normally lasts for one or two hours, with readings taken every five minutes for all parameters. The need for reading accuracy is the same as for a full performance test. There is also a need for a period before the test for setting the steam conditions and ensuring that flow and condition irregularities are eliminated to within the tolerance bands specified for a full code test.

In the case of testing a high-pressure section with a control stage, there is a need to ensure that control valve positions are repeatable because the control stage is a variable pressure design. This is necessary so the test can be repeated with a similar level of pressure drop in the inlet portion of the unit. It is normal for these tests to be undertaken with the high-pressure section operating at the valve wide-open position, or if the unit is load limited due to some mechanical condition, with one or more valves closed and the remainder fully open. However to make a comparison with previous data, such data must exist and have been taken with similar load restrictions in place.

The test instrumentation setup is shown in Figure 4–4. This turbine has a high, reheat, and double-flow low-pressure sections with instrumentation on the high and reheat section at both inlet and discharge from these sections. These positions are shown as *1, 2, 3* and *4*. Because after entering the low-pressure section the steam expands into the moisture region, it is not possible to measure the efficiency of this portion of the unit. In the moisture region, the temperatures measured are those associated with the saturation condition of the steam. That is, there is no indication available of the dryness of the two-phase flow.

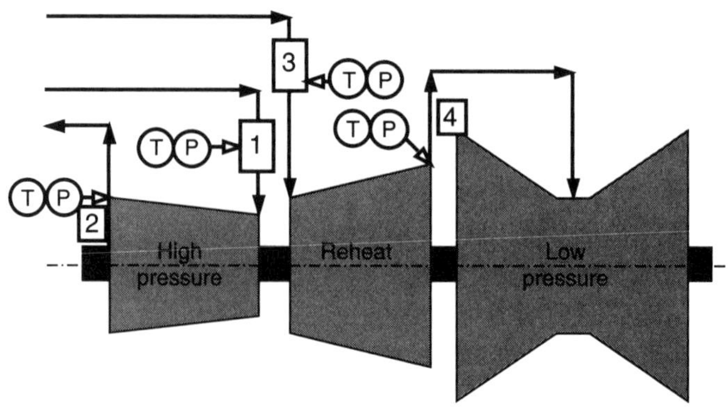

Fig. 4–4 Test Measurements Required to Conduct an Enthalpy Drop Test

Figure 4–5 shows a portion of a Mollier diagram and indicates the variation of steam conditions at inlet to a turbine with nozzle control, or partial arc admission. Here the initial steam conditions as delivered ahead of the steam turbine valves (stop and control) are shown as *Pd* and *Td*. The steam passes through the valves, and during this passage the steam suffers a pressure drop. Because there is no work done by the steam, the expansion occurs at constant enthalpy, from condition *D* to *A*. At condition *A*, the steam enters the first row of stationary blades, and expands through this row to pressure *Pb* at stage discharge. At this pressure, the steam has a temperature *Tb*. This expanded condition is shown on the Mollier diagram as condition *B*. The steam then enters the remaining stages in the high-pressure section and expands to exhaust.

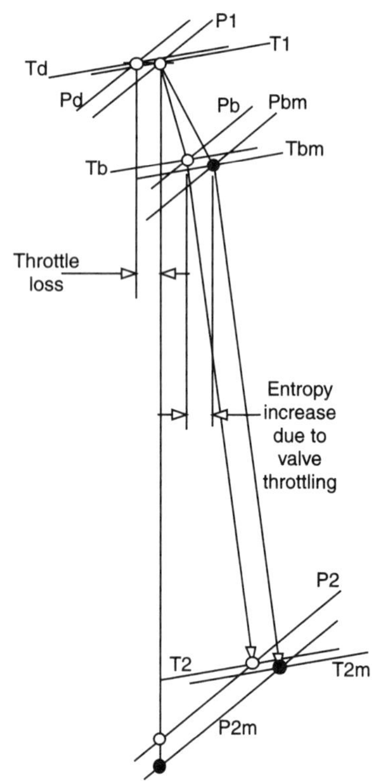

Fig. 4–5 The Expansion Line for a Unit with a Nozzle Control Stage

This condition *B* is reached when the valves are fully open, and there is a minimum pressure drop in the valve system. At part loads the total valves are not open, and there is a loss due to the throttling pressure drop and the partial arc that is available for the admission of the steam. At full load, the end point of the expansion end point in the control stage is shown as *B*. When there are losses in the control stage, the additional losses cause the end point to shift to condition *C*. At this point the steam pressure is shown as *Pbm* at a temperature *Tbm*.

Because of the variation in the steam conditions at discharge from the control stage, it is essential the control valves are in a position that will ensure repeatability of expansion within the valve system.

An improved means, in terms of establishing the state-line efficiency of the high-pressure section, is to have an additional tap point in the bowl at discharge from the control stage. Such a measurement will give a considerably more accurate reading of the efficiency of the stages after the control stage. Unfortunately the control stage is one that is subject to damage due to both foreign particle impact as a consequence of debris carried in from the boiler and also due to solid particle erosion. Therefore, it is important for the control stage to be monitored. If both conditions ahead of the valves and bowl are measured, then it will be possible to obtain an accurate assessment of deterioration in the control stage.

When measuring values at valve inlet and control stage discharge, these values will not give the efficiency of the control stage because these measured values will define an efficiency that will include the energy reduction due to throttling loss in the valves. However, the measured changes in efficiency will be indicative of any change in control stage efficiency. It is possible there could be small efficiency level changes due to flow through the valves, but this is unlikely to have a significant impact on the measured values of efficiency. Before these losses become significant, it is probable mechanical considerations will dictate remedial corrective action for reliability considerations.

Using the results of EDTs

The results of an EDT are used to ascertain the existing condition of the turbine sections which are operating in the superheated region. These superheated steam sections are condition measured and the calculated efficiency values are used to compare degradation from previously established efficiencies from similar code tests. It is normal practice that when a unit is to be removed from service for a major outage for an EDT to be conducted on the superheated steam sections. Then after the unit has been opened, cleaned, and replacement parts installed for those that have deteriorated, the unit is returned to service and the test repeated. From the change in measured efficiency, the predicted improvement in both output and heat rate can be determined. Also the fuel costs savings can be computed for the improvements in the sections where efficiency improvement can be established.

In establishing the total improvement, which is contributed to by the low-pressure section also, it will be necessary to make some assumptions as to the level of improvement in the low-pressure sections. If the initial steam conditions are held constant, then rational judgment can be made of the total improvement from measurements of the total output.

Shown as Figure 4–6 is a typical curve of measured state-line efficiency of a section plotted as a function of time for a unit that is given an EDT every six months. The effect of improving the state-line efficiency through maintenance at a major outage is shown when the unit was opened after five years. The level of efficiency restoration is a function of the condition of the steam path. If the blades are scale coated, it is possible to remove this scale by blast cleaning, but if the blade vanes have suffered damage due to solid particle impact, then only a portion of the loss can be recovered. Similarly, the decisions made concerning seals will have an impact upon the level of efficiency recovery. If seals are replaceable and are damaged, a decision should be made based upon the amount of efficiency that is recovered by replacement. If seals are formed as an integral part of the steam path that it would be difficult to replace, some improvement in seal effectiveness may be possible by dressing.

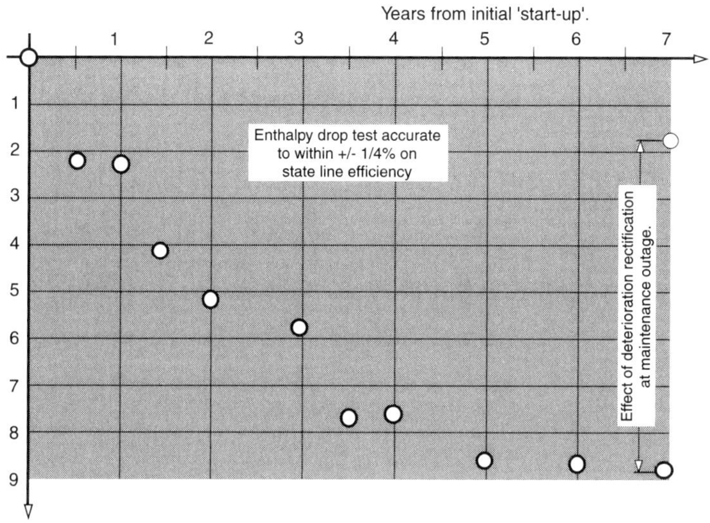

Fig. 4–6 The Results of Enthalpy Drop Tests on a 530,000kW Unit as a Function of Time

OTHER TEST CODES

If an owner intends for a test to be conducted on a new unit for guarantee purposes, it is essential this be detailed in the specification when it is initially issued for competitive bids. Alternately if a test is specified as probable with a decision delayed until after award and at the purchasers discretion, the turbine-generator supplier should be advised as soon as possible after a decision to test is made, because there are certain construction details of the turbine, generator, and the entire plant that need to be factored into the manufacturing schedule to allow for the connection of test instrumentation. In a situation of uncertainty concerning the decision to run a test, bidders will normally specify a time limit for this decision, as this will have an impact upon the engineering design and layout schedule. The requirement to test a unit will involve certain additional instrumentation be supplied and this involves additional cost. Therefore, the longer the decision to test is delayed, the greater the costs involved because there could be a need to modify existing designs and products to achieve test capability.

Most important for the owner is the need to identify the code to which the unit will be tested. There are various national and international codes used for steam turbine-generators, and while the codes vary in detail, their objectives are essentially the same. It will be necessary to identify the code to which the unit will be tested. Common codes include the ASNI/ASME PTC-6 Power Test Code for Steam Turbines, the British Standard Test Code for Acceptance of Steam Turbines, and the Thermal Acceptance Test of Steam Turbines from DIN 1943–1975.

Most commonly the code specified is either the national code of the country of the owner or that of the supplier. If it is the purchaser's choice, the selection should be the code that will provide the best and most accurate information for the particular application.

While these various test codes are intended to achieve measured values of efficiency sufficient for guarantee purposes, there can be considerable differences between them, especially the test margins they recognize and the levels of accuracy they will achieve. These differences exist in terms of the actual test procedures, the extent and quality of instrumentation used, and the levels of correction allowed in the interpretation of test results. These procedural and other differences result in differences in the calculated value of efficiency, and therefore in the adequacy of the test.

In this chapter the ASME PTC-6 code will be referred to. However, the considerations as applied to other codes are sufficiently similar they can be used without too much difficulty.

THE TEST SCHEDULE AND ADVANCED PLANNING

If a test is to be performed on a new turbine-generator set to an acceptable code, and both the manufacturer and owner are to be involved in this, even if one group is involved only as observers, an advanced planning and scheduling meeting should be held as soon as possible after the contract is let. This meeting will allow the objectives of the test to be clearly defined and the aims and responsibilities of both parties to the test to be clearly established. These requirements while quite obvious must be defined and recorded so there is a clear understanding between the parties. Neglecting this requirement can lead to confusion and disagreement as the time for test approaches and can lead to delays. Ultimately this could result in not achieving the requirements of the parties to the test.

As the design of the turbine-generator and thermal cycle progress, flow diagrams for the cycle systems become available. To undertake a test, there are many test instrument connections that are not part of a normal cycle supply but that are required for undertaking a full performance test. The requirements of these connections should be identified on the flow diagrams. Then, as manufacturing drawings are prepared it will be necessary to ensure these connections are present, in a position where they are readily accessible, and able to provide the correct level of information required to establish the efficiency (heat rate) of the unit. These connections are required at specific locations and must have both the correct orientation and be of a size they support the connection of the correct form of instrumentation.

An early requirement of the performance tests is to define what constitutes the turbine-generator. There are various pieces of equipment—mechanical and electrical—within the power generation cycle whose acceptable performance is essential to the satisfactory operation of the power producing turbine, but that do not rightly form part of it, and they certainly are not supplied to the plant by the turbine-generator manufacturer.

An important piece of equipment in this category is the BFP and possibly the BFPT, both of which may have been supplied by other manufacturers with separate guarantees. However, these pieces of equipment have a direct impact on the efficiency or heat rate determined for the main unit. It needs to be established between the owner and supplier which component parts of the cycle will be tested and form part of the turbine-generator guarantee. It must also be noted how the performance of these components will be assessed in terms of the total cycle.

While many of these pieces of peripheral or ancillary equipment will have only a small influence on the cycle and turbine-generator heat rate, if they are efficiency or capacity deficient, then they will have an adverse effect on the turbine-generator performance, and this deficiency must be acknowledged and accounted for. These pieces of equipment must be included in the test or the test cannot be completed. Test and calculation methods exist that allow the influence of these pieces of equipment to be determined and their individual efficiency to be factored into the total turbine-generator heat rate. In fact, these pieces of equipment may be instrumented so their efficiency can be determined during the main test.

The system flow diagrams allow definition of the complete cycle and equipment to be tested. To achieve a test that measures only the meaningful equipment, a means of isolating this to be tested equipment must be devised. This is achieved by isolating this equipment to the greatest degree possible and defining the requirements in an isolation schedule. The requirements of a performance test may require the inclusion of special isolating valves that would not otherwise have been included in the equipment.

Following agreement on the test cycle boundaries and definition of the turbine-generator, an instrument schedule is prepared together with a simplified test cycle flow chart, showing instrument location within the cycle.

The instrument schedule lists and provides the following information:

- the instrumented test point number
- the values to be observed and the anticipated range of measurements
- the type of instrument to be used and its range
- the line designation
- the connection designation and drawing number
- details of calibration requirements and conditions under which recalibration will be required

This information may not be available during the early days of planning for the test. However, this information must be added and agreed to by the parties to the test as soon as convenient after it becomes available.

As soon as the preliminary flow diagrams, electrical connections diagrams, and isolation and instrument schedules are complete, copies are made available to both parties to the test with the request that both review the material and make their comments available.

The test conference

When the preliminary information is available and reviewed, a test conference should be held. This conference is to review every aspect of the test, with particular emphasis placed on those items not specifically covered by the power test code. A typical agenda would consist of the following items.

Definition of the test objectives. This should include definition of what is to be tested. Such definition includes those considerations related to peripheral equipment and possibly any special tests the turbine-generator manufacturer wishes to make to evaluate new design concepts and components.

A statement of the load values at which the unit will be tested. This will include the various loads at which tests will be conducted. If a turbine-generator will operate for extensive periods at partial load, a weighted average heat rate may be specified. This definition adds evaluation for part-load operation. Even if the major evaluation is to be placed on the full load operation, the owner will wish to be aware of part-load performance because as this becomes important as the unit ages. Also, the unit can operate at part load for extensive periods even when the unit is new.

The extent of guarantee. The loads that are guaranteed and any weighting to be applied to part-load operation. It is normal for the turbine-generator manufacturer to guarantee only the full load conditions. However, it is possible that part-load evaluation could be included and guaranteed. Under these circumstances, it is normal for the turbine-generator manufacturer to define the part-load valve condition as a valve point rather than defining a load. This is to ensure the heat rate is determined at a point that is not on a valve loop position and therefore difficult to repeat in subsequent tests.

The steam conditions under which the test will be run. This definition will include the initial and reheat steam conditions and also the exhaust pressure.

The allowable deviations from these values of cycle steam conditions. There are recommendations in the majority or codes for the amount of deviation acceptable. However, there could be situations where deviations might reasonably be requested outside these limits. This is particularly so in the case of special cycles such as geothermal units where the steam pressure and temperature cannot be as easily regulated by the owner.

Definition of responsibility for the provision of adequately qualified manpower. During a performance test, there will be a need for suitable engineers and technicians who can be available for the duration of the test to observe instruments and record their readings. This can represent a considerable cost and provision of these skills should be decided early. The participants during a test can have no other responsibilities within the plant during the test period.

Instrument type, location, and any temporary connections. There are various types of instruments that can be used at different locations. These need to be established early in the planning so they can be made available. Normally there is little disagreement concerning the primary instruments. However, the secondary instruments for measuring flow, pressure, and temperature must be agreed upon since this could have a significant impact on the total costs of the test and the number of instrument recorders required. The type of instrument used will also affect the level of accuracy of the test.

Supplementary tests to be run. These can include such tests as condenser and heater leakage tests and the responsibility for these. Such tests can be conducted concurrently with the total performance test, but sufficient instrument readers must be made available to cover this requirement.

Scheduled date for the tests. Dates must be established in terms of the project schedule. However, schedules can slip. The turbine-generator manufacturer is normally more concerned that the test is conducted as soon as possible after the unit is started, rather than the actual date of the test. An integral part of this scheduling is the time after start-up for conducting the EDT. This test is as important as the main performance test because it allows the manufacturer to establish that there has not been undue deterioration of the unit condition between initial start-up and the test.

Responsibility for the determination of unit condition immediately prior to conducting the test. This is a joint responsibility, and the manufacturer will not normally wish to conduct a test on a unit that has sustained damage or has deteriorated due to some other phenomena. Similarly, the purchaser will wish to be sure the efficiency or heat rate determined by the test represents the unit evaluated and purchased.

Steam path deterioration is rapid at initial start-up and it is normal for the designer to allow some margin in his guarantee to account for this. However, the manufacturer wishes to determine the absolute extent of any deterioration as this can be affected by excess carryover from the boiler of scale forming compounds or even mechanical damage within the steam path. These are factors over which the supplier has little or no control. Therefore, if deterioration between start-up and test is excessive, it is normal to delay the performance test until after the warrantee outage, which normally occurs after about one year of operation.

At the test meeting, the extent of acceptable steam-path efficiency deterioration that is acceptable should be established.

Description of each proposed test run. This will define how each test will be run—the time required to ensure the steam conditions and load are stable and that conditions are acceptable for running the test.

Duration of test run and frequency of test readings. There is usually enough precedent for both time and frequency that this does not require too much definition. However, the frequency of readings can be influenced by the type of instrument used at various locations. This can have some small influence.

Calculation of results and application of instrument correction factors. Agreement regarding the method of calculating the results is normally limited to reaching agreement on the program to be used for computation. With modern computation techniques available, results can be determined quickly and accurately. This is important as it is necessary to qualify each test as it is completed.

Instrument calibration is important, and agreement must be reached as to which instruments will be calibrated, and when. It is often necessary to recalibrate some instruments, particularly the main flow meter at completion of the test series. The throat tap nozzle is subject to loss of calibration as a result of deposits. The parties must agree to the circumstances under which recalibration will be required and the level of wander or deposit in the throat tap nozzle that would nullify the value of the tests.

Margins applied to the results. It is normal in a test conducted to the ASME PTC-6 codes that no test margins will be allowed or required. Some codes recommend margins, but there is value to conducting a test which requires none. If such margins are included, their number should be minimized as these add levels of uncertainty to the values of efficiency and heat rate determined, and make the tests of less value to the owner. Margins must not be applied as safety nets for the equipment suppliers.

At the conclusion of this meeting, after agreement is reached on these and other points, minutes are produced and circulated. These minutes become part of the test procedure because they reflect the philosophy under which the tests will be conducted and so form part of the contract. It is important this test meeting be held as soon as possible after award of the contract as there are decisions reached which will affect the design of the components and systems involved in the tests, including the provision of special hardware and piping connections needed to allow the tests to be conducted satisfactorily.

The date of the test series is dependent upon various factors, but there are EDTs that should be conducted as soon as possible after unit start-up. Most manufacturers will normally run an EDT as soon as the unit is commissioned and the performance tests are conducted—preferably not more than two months later. If this test schedule cannot be achieved, many manufacturers prefer to leave the acceptance test until after the first (warrantee) outage, when the internal condition can be ascertained and, if necessary, corrected.

This first outage is normally 12 to 18 months after commissioning. If deterioration is excessive, the unit can be opened for inspection and correction of any damage or other deteriorating condition that may be present.

About six months prior to start-up, when construction is fairly well advanced, the manufacturer and owner should meet at the plant. At that meeting, a review should be made of the pipe work, to ascertain that provisions for the test have been included. It is also necessary to ensure the necessary connections are in place and are to specification. This is also when plans for instrumentation and data taker locations are selected and the instrument groupings selected to utilize fully the number of assistants required.

After this meeting, a final test procedure is prepared and the schedule and methodology for determining heat rate from the collected data finally reviewed and agreed upon.

TEST CYCLE ISOLATION

To undertake a meaningful test, it is necessary to isolate that portion of the turbine-generator unit to be tested so extraneous in- and out-flows of working fluid do not occur. To achieve this isolation, it is necessary to design in special isolation valves adequate to ensure there are no extraneous flows in the cycle during the test. These valves must be identified and clear indication given regarding special schedules of their test position. This document is known as the isolation schedule. The ASME PTC-6 code states, "the isolation of the system should be accomplished so that the difference between the sums of the measured storage and flows in and out of the system will not exceed 0.1 percent of the throttle flow at full load."

The isolation schedule will provide a list of the valves needed to isolate the test cycle. This schedule indicates the number of valves, their location within the cycle, their function, normal operating position (open or closed), the test position, and the method to be used to verify the test position is correct. It is essential all valves used for isolation are in their correct position during the test to ensure there is neither a net inflow or outflow of working fluid to the cycle. Such flows could distort the enthalpy values at any location.

Once the equipment and flows to be isolated have been agreed upon, the methods to accomplish this must be detailed. There are several methods to accomplish this. The most frequent are double valves with a telltale between them, a blank flange in the line, or a temperature indication—such as the pipe on one side of a valve being hot, and on the other side cold—or one valve and an open pipe.

Once the details of cycle isolation are agreed upon and required valves incorporated, an isolation supervisor is often appointed. This supervisor is responsible for cycle isolation during the test. It is common for the isolation supervisor to run an isolation test prior to the performance test. This test allows the change in stored water over a period of time to be measured which indicates whether satisfactory isolation has been achieved. It is normal after a satisfactory test to tag valves to indicate whether they are to be *open* or *closed* during the test. The normal operating position is also indicated.

Examples of flows that can enter the cycle during the test from external equipment includes those from large volume storage vessels and lines connected to other parts of the station system, particularly those that may be interconnecting with other units or systems in the plant. It is necessary for the test design engineer to examine possible sources of such extraneous flow and be sure the isolation system is capable of isolating the unit for the duration of the test.

	Maximum Variation from Design	Maximum Permissible from Design	Instrumentation Uncertainties Test Fluctuation e
A. Electrical Output			
1. Indicating wattmeter	+/- 5.0%	+/- 25%	2.20%
2.Timing wattmeters disc revolutions		+/-0.25%	1.90%
3. Precision wattmeters			0.28%
4. Duplicate precision wattmeters			-.20%
B. Throttle Flow			
1.Recorders			4.00%
2.Manometer or station nozzle			1.70%
3.Manometer and calibrated flow section			0.26%
C. Main Steam Pressure	+/-3.0%	+/-2.0%	
Cold Reheat Pressure			
Hot Reheat Pressure			
I.P. Exhaust Pressure			
1.Transmitter, recorder or bourdon gauge			4.40%
2.One deadweight gauge/measurement			0.28%
3.Two deadweight gauges/measurement			0.20%
D. Throttle Temperature	+/-17%	+/-4.0%	
1.Recorder or indicator with thermocoupler resistance element			0.80%
2.Two thermocouples, with cold junction and semi-precision potentiometer			0.12%
E. Cold Reheat & I.P. Exhaust Temperature			
1.Recorder or indicator with thermocouple or resistance element			0.50%
2.Two thermocouples with cold junction and semi-precision potentiometer			0.05%
F. Hot Reheat Temperature	+/-17%	+/-4.0%	
1.Recorder or indicator with thermocouple or resistance element			0.60%
2.Two thermocouples, with cold junction and semi-precision potentiometer			0.08%
G. L.P. Exhaust Pressure (For Load Correction)	**+/-0.05 psi**		
1.Manometer or absolute pressure gauge	or 2.5%		0.30%
2.Two manometers with basket tubes			0.16%

Table 4–1 Test Instrument Requirements

TEST INSTRUMENTATION AND CALIBRATION

For the performance test to be meaningful, and provide data which will allow an accurate heat rate and output to be established, and also subsequently allow condition monitoring of the unit throughout its operating life to be of value, instrumentation must be selected which will provide a level of accuracy and reading repeatability consistent with the objectives of the test. There are a range of parameters which must be measured, and there are various classes of instrumentation which can be used for each. The instrumentation used for each measurement is selected to provide the level of accuracy required, reflecting the criticality of the parameter, and its possible impact on the final values of performance.

This instrument accuracy requirement applies to subsequent full code tests and EDTs run to monitor deterioration.

The accuracy required of test instrumentation and the level of variation of steam conditions permitted during an ASME PTC-6 test to obtain meaningful results are shown in Table 4–1.

Combined uncertainties

The gauges and instruments used to establish the cycle and unit parameters are permitted a degree of uncertainty as indicated by Table 4–1. The sum effect of these uncertainties on the accuracy of the calculated values can be found from equation 4.3, where *e1.2.3 ... n* represents the individual instrument or reading uncertainties.

$$Ep = \sqrt{\left[e_1^2 + e_2^2 + e_3^2 + e_4 \; . \; . \; . \; . \; . \; . \; . \; . \; e_n^2 \right]} \qquad (4.3)$$

where

Ep	=	the total probable uncertainty
$e_{1 \ldots n}$	=	the individual reading uncertainty ranges
N	=	the number of readings used in the calculation

A full performance test conducted in accordance with the ASME PTC-6 should provide a test accuracy in the order of +/-0.25%. This is considered sufficiently small that no test margin is required by the manufacturer or owner. That is, the calculated value is accepted as being the test efficiency, and it either meets or does not meet the guarantee.

There are various parameters that must be measured around the power cycle to establish its efficiency, and there are primary and secondary values of these measurements at various locations that must be taken during any test. The criticality of the particular parameter will determine the form of instrument used to measure it. There are three main quantities that must be gauged—the level of thermal energy, the working fluid flow quantities, and the power generated. These parameters are measured by various methods.

Measurement of flow quantities

Flow is one of the more important parameters measured in the course of a performance test. This is particularly so in the case of the main working fluid flow because any uncertainty in its measured value represents a one-for-one error in the accuracy of the final determined value of heat rate. However, the flow measurements required in any test can be considered in two parts. The first is the main steam or condensate flow, which varies almost in direct proportion to the power output of the unit. The second is that group of secondary flows—flows required to and from equipment within the cycle. These flows are often part of, or are extracted from, the main flow. These flows have to be considered in determining the total thermal energy available at any location.

An accurate method of measuring flow is to collect and weigh the quantity of water that flows at any point within a given period of time. This method is acceptable. However, there are difficulties associated with the calibration of the holding tanks and the need to be able to contain a large quantity of water. For most large applications this proves too difficult, and a more realistic method is to monitor working fluid flow rate, which is held sensibly constant during the period of the test.

Because of the difficulty measuring the flow rates of steam and the difficulty in calibrating such a measuring device under simulated field conditions, it is preferable, where possible to monitor the condensate or feed water flow with a calibrated flow meter.

Primary flow measuring devices. The quantity of steam or water can be measured by monitoring orifices, nozzles, or venturies. In each case, these devices require careful calibration, which should normally be undertaken just prior to the test. It is important these flow-measuring devices are calibrated at or as close to the actual Reynolds Number of the flowing working fluid as possible. This calibration should be undertaken before the test and, depending upon the extent of deposits on the inner surface, repeated at completion of the test, to ensure there has not been extensive change in the surface condition that would influence the flow characteristics during the test period and therefore the calibration. Such changes can occur if there is any possibility of deterioration or deposits that would affect the flow coefficients of the measuring device.

The flow quantities can be measured by monitoring the steam flow or the flow of condensate after it is extracted from the condenser and passed through any low-pressure heaters. The normal preference for the engineer responsible for conducting the tests is to monitor the condensate flow as there is less uncertainty concerning the measurement of the fluid rather than gaseous flow. This is especially so if the initial flow to the turbine is a two-phase mixture.

There are, however, instances where it is not possible or practicable to measure the condensate, such as in a geothermal application. Under these circumstances, the inlet flow must be monitored. This is done using a calibrated flow orifice.

For the measurement of condensate, a flow meter employing a differential pressure gauge must be used. There are three basic measuring devices that can be considered. These are shown in Table 4–2.

Device.	Advantages.	Disadvantages.
Orifice Plate	Economical to manufacture Size of orifice readily changed	Liable to lose sharpness and so change coefficient if left long in situ
		Little affected by deposits
Venturi Tube	Minimum Pressure drop	Sensitive to Deposits
	Expensive to manufacture	
Throat Tap Nozzle	Capable of high accuracy	Sensitive to deposits
		Expensive to manufacture

Source: ASME PTC-6 Power Test Code for Steam Turbines.

Table 4–2 Differential Pressure Flow Measuring Devices

A typical test orifice is shown in Figure 4–7, which indicates the orifice plate is assembled into a spool piece that can be inserted into the condensate return line at some convenient position. When a code test is run for the condensate return line from the hot well to the feed pump, it is normal for this to be designed to allow assembly with sufficient up and downstream line length so there is no flow directional change that could influence the flow rate readings. For the elliptical throat tap nozzle, the dimensions for flow straightening, according to ASME standards, are as shown in Figure 4–7. Figure 4–8 shows an orifice plate that can also be used for flow quantity measurement.

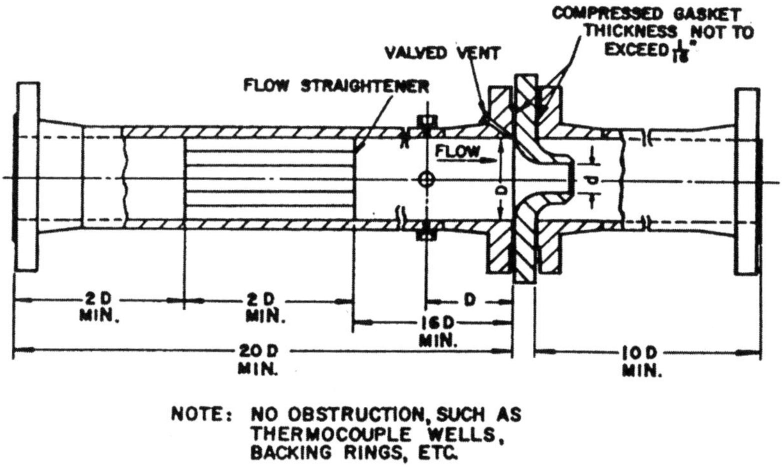

Fig. 4–7 A Test Nozzle Suitable for Measuring Primary Flow

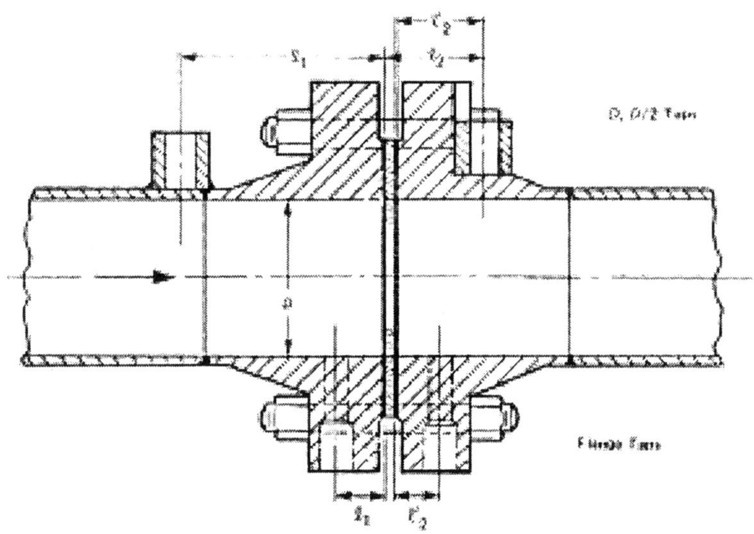

Fig. 4–8 Orifice Plate, Showing the Location of the Plate and Pressure Taps

This type of assembly allows the easy removal of the orifice plate for inspection or change if necessary. Pressure tap points are made available on both the up and downstream flanges. These points are then connected to a differential pressure gauge calibrated to measure flow rate.

Figure 4–9 shows an elliptical throat tap nozzle. (ASNI/ASME) This device, like the orifice, must be installed in a straight run of piping as shown in this figure, and there are minimum straight lengths of pipe up and downstream of the nozzle. These straight lengths are necessary to eliminate flow irregularities in the fluid flow that would distort the accuracy of the readings.

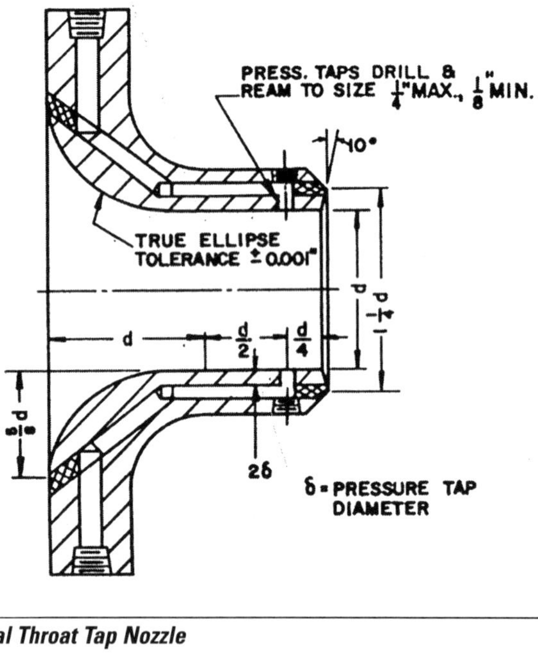

Fig. 4–9 Elliptical Throat Tap Nozzle

These flow quantity measuring devices should preferably be connected into the condensate flow line in a removable loop. This loop pipe, containing the flow measuring device, provides an alternative line through which the condensate can be passed and it must be capable of isolation so it can be mounted or removed from the system without the need to shut down the unit. This measuring device should also be mounted in a horizontal run of pipe. This horizontal requirement is necessary to eliminate any differences introduced into the readings by the difference in pressure head that will exist in the fluid.

If the flow measuring device is to be used in any other than a horizontal configuration, it should be calibrated in this position to account for any differences in the water density in the vertical run. During the calibration of the flow measuring device it is necessary to establish the correction factor for flow as a function of the flow Reynolds Number. A typical correction curve is shown in Figure 4–10. Under ideal conditions the nozzle should be used in the region where the flow coefficient is nearly constant. The calibration of a flow measuring device is shown in Figure 4–11.

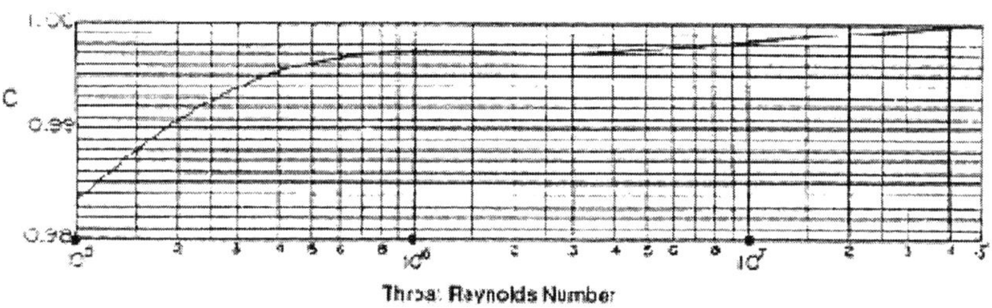

Fig. 4–10 The Calibration Curve for a Flow Measuring Section as Established by a Calibration Test

Fig. 4–11 The calibration of a flow measuring device.

Below a throat Reynolds Number of two million there is flow transition from the lamellar to turbulent regime in the boundary layer. For this reason the transition range should be determined during the calibration process and avoided during the performance test.

Secondary flow measurement. While the measurement of primary flow is critical to the accuracy of the test, there are a number of secondary flows that need to be determined to allow the total energy distribution around the cycle to be established. The method of establishing these secondary flows depends upon the cycle design, the conditions at the measuring points, and their location. The actual method to be used at any location should be evaluated and agreed between the parties to the test. A recommendation of the British standard code is that the total uncertainty introduced by the methods used to establish the secondary flows should not introduce an uncertainty in the test heat rate in excess of +/-0.10%. (British Standard Test Code 1974)

In addition to the options available for direct measurement of the secondary flows, it is possible to calculate them by performing a heat balance around the section or apparatus concerned. Typical of the secondary flows that would be measured by an orifice and can also be established by calculation are those extracted from the turbine for regenerative feed heating.

If the flows are established by calculation, it is essential the measurement of the steam/water parameters be determined by precision calibrated instrumentation that meets the accuracy requirements of the test and does not adversely affect the accuracy of the calculated heat rate.

If there is a possibility that flow is to be measured and this is a two-phase fluid, precautions should be taken to ensure an accuracy of 0.50% is achieved. Tracer techniques are now common and of considerable accuracy for this type of flow.

The major secondary flows to be considered are those associated with steam extraction for regenerative feed heating, steam extracted from the unit above the reheat point, and others such as gland leakage steam and steam to the air ejectors. The actual measurement requirements around a feed heater are dependent on the disposition of the heater drains, but often these quantities can be determined with acceptable accuracy by the combined use of measuring devices and heat balance calculations.

In the case of nuclear units, it is necessary to monitor and measure the drains from moisture separator and live steam reheater vessels. The most satisfactory methods in these instances employ the use of tracer techniques.

Measurement of power output

The most significant power measurement is the determination of the electrical power output from the generator or the mechanical power developed in a mechanical drive turbine. There are, however, other levels of power consumption within the unit and cycle that must be determined within reasonable limits of accuracy so the net power generated and the actual efficiency can be established.

Generator electrical power output. The net electrical power developed within the unit is defined as the power produced at the generator terminals minus any power supplied to the generator exciter together with any power supplied to other equipment or service requirements of the cycle.

The actual power at the generator terminals is best determined by the use of wattmeters, and either integrating or indicating instruments can be used. There are two methods of connecting these instruments that can be used for a three-phase alternating current (ac) generator.

- The first method suitable for a three-phase unit, in which the neutral is either grounded or not, requires that a separate wattmeter be connected within the three phases. For three wattmeters as shown in Figure 4–12, the wattmeters are connected through suitable current transformers, one to each line and PTs as shown from each phase to neutral. It is necessary to have both the current and voltage transformers calibrated.

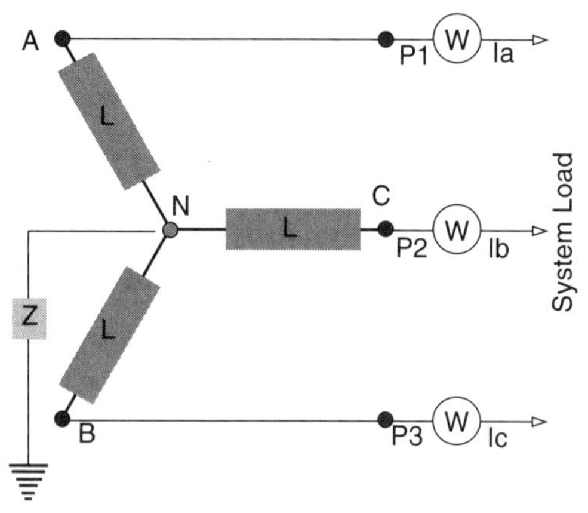

Fig. 4–12 Power Measured by Three Wattmeters

In this arrangement, the total power developed is the algebraic sum of the three wattmeter readings. The total power *ΣW* can be found from:

$$\Sigma W = Wa + Wb + Wc$$

This is the total instantaneous power in the three lines.

- The second method can be used when the neutral is not grounded, and it requires the use of two wattmeters. These are connected as shown in Figure 4–13, which is a *Y* connection. In this case the instantaneous power measured is

Wattmeter	Wa: current = Ia, and voltage is Va - Vc
Wattmeter	Wb: current = Ib, and voltage is Vb - Vc
Total power	= Wa + Wb = Va.Ia + Vb.Ib + Vc.Ic

This again represents the total instantaneous power in the three lines.

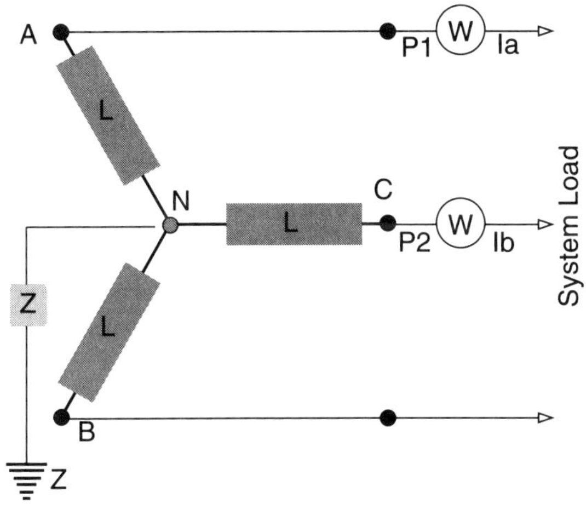

Fig. 4–13 Power Measurement by Two Wattmeters

Similarly for the delta connection in Figure 4–14, the power is measurable as the sum of the two instantaneous powers *Wa* and *Wb*.

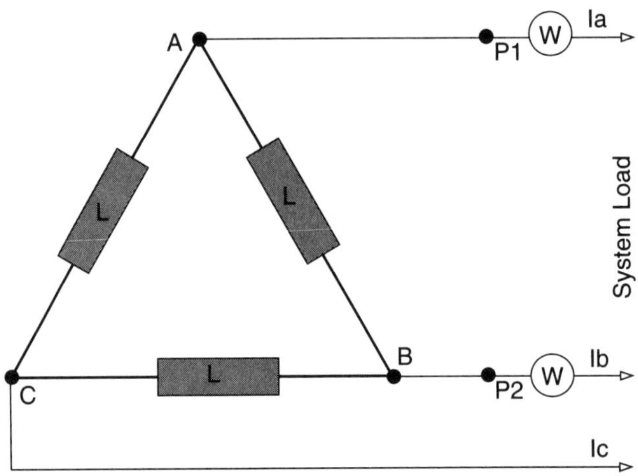

Fig. 4–14 Power Measurement by the Two Wattmeter Method for the Delta Connected Circuit

Integrating wattmeters are most often used, as these provide a value of power integrated over the period of the test. These meters require calibration before the test, and calibration correction factors should be available at completion.

Power factor. The power factor Φ can be established from the ratio of the wattmeter readings using equation 4.4.

$$\Phi = \sqrt{3}\,\frac{Wa - Wb}{Wa + Wb} \qquad (4.4)$$

Measurement of exciter power. If the exciter power is supplied from the main generator, this must be measured separately so the quantity used can be subtracted from the total output. Or the power output can be measured beyond the exciter connection so the net power is determined.

Measurement of mechanical drive power. The mechanical power output of a main drive or auxiliary turbine is normally measured by a dynamometer or torque meter. These measurements must be made by calibrated instruments.

The boiler feed pump power

In the event it is required to establish the BFP and BFPT efficiency and power consumption with a degree of accuracy that would make it acceptable to distinguish between its contribution to the cycle efficiency and that of the main unit, there is a means of doing this. If the BFPT exhausts directly into the condenser, making its exhaust steam conditions wet and therefore difficult to establish with an acceptable degree of accuracy the expansion line end point, then it becomes necessary to devise a method of calculating its effect on the total cycle. This can be done in the following manner:

- Establish the water horsepower of the boiler feed pump by means of the pressure rise through the pump. It is not possible to take account of the small temperature rise through the pump as this rise is small and will have a relatively minimal effect on the accuracy of the final readings.
- Using the pump characteristic curves provided by the pump supplier, and possibly witnessed at a separate test, the value of the mechanical input corresponding to the water horsepower (WHP) can be determined. This pump input is the power extracted from the turbine.
- Using a calibrated nozzle, the steam flow m to the BFPT can be determined.
- The pressure and temperature at inlet to—$P1$ and $T1$—and outlet from—$P2$ and $T2$—to the BFPT will be measured.

$$\text{Efficiency of the BFPT } (\eta\text{bfpt}) = \frac{\text{Turbine Mechanical Output}}{\text{Heat Input to Turbine}}$$

$$\text{Therefore, } \eta\text{bfpt} = \frac{\dfrac{\text{Water Horse Power}}{\text{Boiler Feed Pump Efficiency}}}{\text{(Steam Mass FLow) x (Enthalpy In) - (Enthalpy Out)}}$$

$$= \frac{\text{WHP}/\eta\text{bfp}}{\text{m . (H1 - H2)}}$$

Iteration from an assumption concerning the turbine efficiency *ηbfp* will establish the expansion line endpoint, and allow the endpoint enthalpy *H2* at *P2* to be determined. Using this value of *H2*, the equation can be checked for consistency. If these do not agree, then another value of the state-line efficiency will be assumed. This procedure will be continued until agreement is reached and the BFPT efficiency determined.

Measurement of thermal energy

Thermal energy is defined as the enthalpy level of the steam. However, it is not possible to measure enthalpy directly, but this quantity can be established in terms of pressure, temperature, specific volume and/or entropy, or any two of these parameters, knowing two such values the enthalpy can be determined from the steam tables. Unfortunately entropy is not a measurable characteristic, and specific volume would be difficult to determine. However, enthalpy can be established in a performance test by direct measurements of steam pressure and temperature at a common point. Therefore, for testing, methods must be developed for establishing these values of pressure and temperature with extreme accuracy.

Measurement of pressure. The pressure measurements made as part of the performance test can vary in magnitude from the high initial pressure at inlet to the turbine to sub-atmospheric values at the condenser. The requirements for these two extremes are different, and the most suitable instrumentation for any application is dependent upon the pressure range to be measured. For high-pressure points there is a need to use gauges that retain their calibration when operating at high temperatures and can, if necessary, be checked to national standards. Accurate measurement of static and differential pressure can be made with modern pressure transducers to an accuracy of 0.10%. The *ASME Performance Test Code Instrument and Apparatus Supplement PTC19.2 Pressure Measurement* is available to provide guidance in making this measurement.

The form and size of a pressure tap point and its location within the system are important. These factors can have an influence on the validity and accuracy of the measured values. For measurement of static pressures, the tap point location and size are not as important, since the fluid is relatively static and no large uncertainties are introduced by velocity effects. However, where differential pressures are being measured and are affected by relative flow velocities, the location and form factors can introduce an unacceptable level of uncertainty if their form is not known and able to be accounted for.

For pressures above about 35 psia, a pressure transducer of 0.10% should be used. Alternately, calibrated dead weight gauges would be suitable. When the measured pressure is predicted to be below about 35 psia, a calibrated manometer should be used. The major pressures at connections to turbine sections and other flow points should also be measured by duplicate gauges, which must be located in locations which are known to minimize measurement uncertainties. If transducers are used for measuring differential pressures, a class with an accuracy of 0.005% is required. Deadweight gauges should also have duplicated readings and be calibrated before use.

If Bourdon gauges are used, they should be isolated from the working fluid to ensure that they do not become influenced by its temperature, which can affect the metal temperature and expansion characteristics.

An important pressure measurement is that used to measure differential pressure on main steam flow. These manometers are arranged so they can by means of verniers be read to accuracies of +/-0.01.

Atmospheric pressure should be determined by a precision-type manometer. Values of atmospheric pressure are required to correct manometer readings, which measure many important pressure readings at the lower pressure levels, including the turbine exhaust pressure.

Turbine exhaust pressure for condensing turbines should be measured using pressure transducers. Absolute pressure gauges can also be used and are suitable for turbine exhaust pressure because there is a considerable amount of energy associated with small pressure level changes in the low-pressure range. That is, the effective exhaust pressure has a considerable effect on the energy available from the steam for energy conversion. To establish this pressure as accurately as possible, it is necessary to separate the velocity head from the total pressure to establish the pressure head. This is achieved most accurately by the use of basket tips that surround the pressure sensors and eliminate the velocity effect to the greatest degree possible. It is important to use the correct number of pressure sensors and place them in the exhaust annulus to achieve a meaningful measurement as there can be a considerable difference of velocity and pressure within the exhaust hood. There are requirements for the number and location of these pressure gauges—as defined in the test code. The pressure used should be the average of the gauges.

Measurement of temperature. Steam temperatures are the second parameter that must be measured to establish the enthalpy levels of the working fluid. Guidance is given for temperature measurement in ASME *PTC 19.3-1974.*

The calibrated thermocouple, with continuous wire from a hot to cold junction is the major source of temperature measurement in any tests undertaken to establish steam turbine performance. Possibly the most common metal combination in present use is the chrom-constantan used with the cold junction in a thermos flask, whose temperature is maintained at 32° F by the use of ice. Each ice point should be electrically insulated.

Both pressure and temperature measuring devices should be calibrated to a national standard before performance guarantee tests are undertaken. Thermocouples are inserted into the steam or water flow using wells that are weld attached to the lines. These wells are constructed to be as thin as possible consistent with the internal pressure and the stresses induced by the steam flow. The wells are also designed so the thermocouple does not have excessive clearance on its diameter and bottoms on the well to ensure adequate heat transfer from the fluid through the walls to the thermocouple.

Measurement of steam quality

The determination of steam quality has always been a difficult measurement to make and has normally been completed by throttling calorimeters and the measuring drain flow. Another method has been developed for use in steam turbines operating on wet steam where it is necessary to establish quality with a degree of accuracy sufficient to support the determination of cycle and unit heat rate with the level of accuracy required to comply with the requirements of a code test. A method being developed but not yet refined to the extent it will be able to be used for performance testing, is the use of a laser probe capable of establishing moisture content as a function of blade radial position and then integrate the individual readings over the total length to establish moisture content. These methods of establishing steam quality can be considered:

Calorimeter measuring method. The calorimeter used for steam quality measurement uses a sample of steam extracted by a probe from a steam/water line. The difficulty with this method is in determining that the sample extracted is representative of the total steam flow.

Measurement of drain flow and heat balance. The quality of steam in an extraction line can be established by measuring the pressure at various points around the heater, and there is a requirement to measure the drain flows into and out of the heater. With these flows and pressure readings, a heat balance can be performed around the heater sufficient to allow the condition to be calculated.

Tracer injection techniques. Radioactive and non-radioactive tracers can be injected into the water/steam mixture flow to determine the moisture content of units operating in the moisture region for nuclear units operating on steam generated in the water cooled reactors. (Herzog, Cotton, and Schofield 1972)

Laser probe traverse method. A recent development in establishing moisture content is the traversing of a blade row in the radial direction by a probe that counts moisture drops and measures their size and integrates these droplets into a moisture content for the two phase flow.

This procedure has not yet been accepted as a suitable method for any code, but as the technique improves in terms of accuracy, it is anticipated it will be given consideration, and eventually incorporated as an acceptable measuring tool.

Measurement of time

The time measurements taken during a test are used either to establish the interval between readings or where integrating meters are used to time the complete test. The time interval of the test periods and the observations may be determined by signals from a master clock. This timing device should be an accurate quartz clock or else a timekeeper should be used to give signals. Alternately, observations may be timed by watches of the individual observers. If individual watches are used, then these should be synchronized at the beginning of the test. It is also possible to use a light signaling system, where observers are advised to record values as a light flashes.

Measurement of unit speed

In the event unit speed is to be measured, a revolution counter can be used over a time measure, or speed can be measured by means of a stroboscope. If the unit is synchronized to a major distribution system, then the frequency of the system will control unit speed to accurate levels.

SIMPLIFIED ASME PTC-6 TEST

The cost of conducting a full test to the ASME PTC-6 code is expensive and owners are reluctant to conduct them despite the advantage such tests provide for the life of the unit. The costs associated with such a test will often be recovered within a few years from diagnostic analysis of subsequent tests and the advantages this can provide in terms of fuel costs.

To lower the costs of conducting a performance test without sacrificing test accuracy to the extent the test becomes meaningless, a method was developed utilizing an accurate high-pressure flow meter that could be incorporated between the top heater and the boiler. (ASME 1970) (Bornstein and Cotton 1981)

This location reduces the need to monitor and establish secondary flows to heaters to the extent accurate flows can be established by performing a heat balance around them. In units which employ reheat it is necessary to conduct a balance around any heaters above the reheat point, so as to determine the flows that occur there so the quantity of steam being reheated can be established with accuracy.

This simplified test, while providing sufficient information to determine the cycle heat rate, will not provide as much diagnostic information concerning peripheral equipment as a full test. If there are problems within the cycle that tend to degrade the cycle efficiency, these tests will not provide as much information to help identify the need for corrective action.

The flow nozzle used for this test is located above the boiler feed pump in the high-pressure portion of the pipe and must be suitable for these higher operating pressures. Also, the fluid will be at a higher temperature than is normally experienced by flow meters in turbine tests. The flow meter shown in Figure 4–4 has an inspection port produced in the sidewall that allows the flow section to be examined for deposits prior to any tests. A section through the inspection port is shown in Figure 4–15.

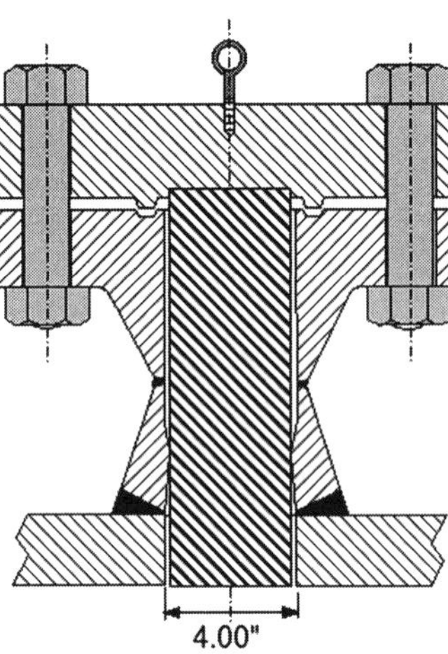

Fig. 4–15 Detail of the Inspection Port

This test does not provide the same level of accuracy as the full test but can achieve a +/-0.38% level.

REFERENCES

ASME PTC 65 Report–1970, "Simplified Procedure for Routine Performance Tests of Steam Turbines."

ASME MFC-3M-1989. "Measurement of Fluid Flow in Pipes Using Orifice, Nozzle, and Venturi."

ASNI/ASME PTC-6 *Power Test Code for Steam Turbines.*

Bornstein, B. and K.C. Cotton. "A Simplified ASME Acceptance Test Procedure for Steam Turbines." *Combustion.* March 1981.

BS 752–1974. *British Standard Test Code for Acceptance of Steam Turbines.*

Cotton, K.C., et al. "Radioactive Tracer Techniques for Testing Steam Turbines in Nuclear Power Plants." General Electric Publication GER-2631.

Crimm, H.G. Jr. and J.C.Westcott. "Turbine Cycle Test Procedure at Potomac Electric Company." *Combustion,* March 1980.

DIN 1943-1975. *Thermal Acceptance Test of Steam Turbines.*

Herzog, W.G, K.C. Cotton, and P. Schofield: "ASME Steam Turbine Code Test using Radioactive Tracers." The ASME Winter Annual Meeting, N.Y., November 1972.

James, J.W. "Performance Acceptance Testing of Steam-Turbine-Generators." Canadian Electrical Association, Thermal Power Section, Winnipeg, Manitoba, March 1967.

Neustaedter, R.H. and W.P. Sanders. "An Outline of the Requirements for Performance Acceptance Testing of Steam Turbine-Generators in accordance with ASME." PTC-6, 1964. Canadian General Electric Publication, March 1969.

Appendix

VOLUME IIIB

Foreword

Preface

Acknowledgments

Author's Note

5 Mechanical Design Considerations of the High and Low-Pressure Casings

Introduction

Components Comprising the Turbine Casing

Functions of the Shells or Casings

The high-pressure/high-temperature sections

The low-pressure/low-temperature sections

High-Pressure/High-Temperature Casings

Pressure Staging and Multiple Shells

The Low-Pressure Casings

Low-Pressure Casing Arrangement

Low-Pressure Casing Structures

The outer upper shell

The outer lower shell

The inner casing

Cast Low-Pressure Sections

The material specifications

Acceptance level of casting defects

Thermal Gradient and High-Pressure Shell Design

Estimating Low Cycle Fatigue Life

Thermal Gradient in the Low-Pressure Inner Casing

High-Pressure Turbine Shell Materials

Steel Plate for Low-Pressure Casing

The plate specifications

Plate faults

The major internal defects

The major surface defects

Casing Weldment Considerations

Casing Weldment Details

Structural welding faults

Welding Consumables

High-Pressure Shell Manufacture

Shell Casting Faults

Shrinkage cavities

Cold cracks

Hot tears

Porosity

Scabs

The Upgrading of Castings

Shell Manufacturing Tolerances

The Joining of Casings Parts

The bolted horizontal joint

The shrink ring joint

The end loaded rotor design (barrel construction)

Threaded Components

Forms of pitch errors

Bolt stresses

The types of failure in threaded components

Stress relaxation

Preloading of bolts

Reuse of high strength bolts and studs

The tightening of large bolt and stud elements

Threaded component material

Pipe Connection Points on the Casings

Bolted connection

Upper half weld

Lower half weld

Slip-Ring Connection

Steam Inlet and Nozzle Box Systems

Explosion or Relief Diaphragms

High-Pressure Packing Heads

Casing Exhaust Geometries

References

6 Mechanical Design Considerations of The Stationary Blade Rows

Introduction

Function of The Stationary Blades and Diaphragms

Diaphragm Construction and Manufactured Types

Stationary blades for direct mounting in the casing

Diaphragms for locating in an inner or outer casing

Form of the Diaphragms

Pinned construction

Welded construction

Cast construction

Fabricated by weldment

Machined from solid blanks

Steam Inlet Nozzle Boxes and Plates

Nozzle plate construction

Nozzle blocks

Nozzle plate attachment to the steam inlet chamber

Inlet region and stress concentration

Nozzle plate materials

Built-Up Stationary Blade Rows

The Stationary Blade Vane

The Stationary Blade Row Vane Form

Two-dimensional considerations

Three-dimensional considerations

Vane tilt

Profile details

Stationary Blade Row Dimensional Requirements

Stationary Blade Annulus Finishing and Tolerances

The diaphragms

The Diaphragm Horizontal Joint

Steam path blading continuity

Error types and corrective methods

Adjustment after vane correction

Vane Cross-Section Irregularities

Nozzle box

Extended vanes (for axial strength)

Throttle controlled units at their horizontal joint

Form of the Diaphragm Rings or Webs—The Secondary Functions

The provision of moisture removal capability

The provision of steam sealing capability

The provision of steam guiding facilities

Steam seal face

Diaphragm Sidewalls

Diaphragm Location Within The Casing

Diaphragm Stresses And Material Considerations

Material considerations

Stress considerations

References

7 Mechanical Design Considerations of The Turbine Rotors

Introduction

The Rotor

Functions of the rotor

Rotor Form

Rotor material development

Rotor Construction and Manufacturing Options

Monobloc (shrunk assembly)

Built-up (shrunk assembly)

Turbine rotor discs

Welded (a welded assembly)

The hybrid rotors

Heat Indication Tests

The thermal stability test

Operating Instability Due To Rubs (Newkirk Effect)

Rotor and Wheel Stresses

Types of rotor stress

Rotor Temperature Control

Rotor pre-warming

Typical pre-warming instructions

Operating temperature gradients and control

Rotor Geometry and Machining

Section changes and rotor span

Seal constrictions

Wheel to central spindle transition radii

Blade attachment regions

Component attachment areas on the rotor

The Critical Speed of Rotors

The criticals

Significance of critical speeds

Speed vibration curves

Rigidly coupled rotors

Values of critical speed

Vibration of Turbine Generator Rotors

Causes of unbalance

Rotor Overspeed Testing

References

8 The Rotating Blade Rows Mechanical Design Considerations

Introduction

Blade terminology

The Blade Vanes

Vane types (by design philosophy)

Impulse and reaction profiles

Blade Vane Requirements and Design

The Requirements of the Blade Vane

Profile and Cascade Quality Requirements

Blade profile definitions and tolerances

Blade cascade definitions and tolerances

Passage swallowing capacity

Blade Profile Types

The forms of profile

Two Dimensional Vane Profile Design

Determination of profiles

Profile Variation With Vane Height

Predicted Expansion Passage Flow Patterns

Operating Stresses In The Blade Vane

Methods of Stress Determination

Direct centrifugal stress

Example 8.1

The effect of the tie wire holes

For the outer tie wire hole position

For the inner tie wire hole position

Example 8.2

Solution

Steam bending stresses

Pressure bending stresses

Example 8.3

Solution

Centrifugal bending effects

Determination of centrifugal bending effects (offset stress)

Blade vibratory stresses

Modifications To The Simple Stress Diagram

References

9 The Rotating Blade Row Root Attachments

Introduction

Forms of The Root

Axial entry

Radial entry

Tangential entry

The internal type root

The straddle type root

The weld-attached blade

Blade Vane Placement On the Root Platform

Tangential Entry Blade Root Access Window

Root Load Transfer Surfaces

Root Side Grips

Blade Root Stresses

Axial entry type

Radial entry type

Tangential entry type

Finite element analysis (FEA)

Considerations of Load Bearing Surface Pressure

Unequal Load Sharing Within Blade Roots

References

10 Mechanical Design Considerations Of Stage Hardware, Coverbands, and Tie Wires

Introduction

Functions Of the Stage Hardware

Functions of the coverband

Functions of the tie wires

Batching Of Blade Ties

Single tie elements

Forms of the Tie Wire

Wire cross-sections

Continuous wires

Wire tangential clamping

Connecting wire ends

The integral wire

Staggered wires

Weld Grouping of Blades

Forms of the Coverband

Type A (plain cover, no seal)

Type B (axial seal)

Type C (radial seal)

Type D (radial seal platform)

Type E (axial seal and radial seal platform)

Type F (special purpose bands)

Forms of the Integral Coverband

- Pre-twisted blades

Shaping of the Coverband Ends

Tie Wire Manufacture and Assembly

- Continuous wires
- Integral wires
- Manufacturing and assembly tolerances

Braze Attachment Of Tie Wires

- Braze attachment procedure
- Brazed tie wires

Tenons and Coverband Attachment By Riveting

- Forms of the tenons
- Tenon and hole requirements
- The rivet
- The riveting process
- The fox-holed tenon
- The rolled rivets
- Mounting the coverband
- Tenon preheat
- At completion of riveting
- Shroud band shaping

Coverband Removal and Reassembly

- Removal of an existing coverband (band to be scrapped)
- Removal of an existing coverband (band to be saved)

Coverband Manufacture

- Manufacturing tolerances

Tie-Wire Stresses

- Continuous wire stresses
- Integral snubbers on the vane

Coverband Stresses

The coverband bending stresses

References

11 Rotating Blade Manufacturing Methods, Materials, Spatial, and Quality Program Requirements

Introduction

Blade Manufacturing Cost Considerations

The Blade Manufacturing Processes

Basic form production by material deformation

Basic form production by metal cutting

Basic form production by electric discharge machining

Profile and Cascade Tolerances

Blade profile

Gauging the profile

Blade cascade

Profile and Cascade Errors

Secondary consequences

Passage Swallowing Capacity

Blade Vane-to-Root Alignment

Special Processes Applied to the Vane

Blade Attachment System Manufacture

The side grip

Production and placement of load bearing surface radius

Surface Finish Requirements

Blade vane requirements

Metal surface finish specifications

Comparisons of surface specification

Requirements To Accommodate Stage Hardware

Blade Materials

The forms of blade steel

The Manufacture and Form Of Blade Steel

The material specification

The requirement For material test certificates

Bar stock material

Forged blade material

The Alloying Elements In Blade Steel

The Mechanical Properties of Turbine Blade Materials

The Heat Treatment of Alloy Steel

Typical Specifications For Blading Steel

Titanium Alloys For Blades

Quality Program Requirements and Surveillance

The engineering review

The quality assurance program

Inspection plan

Product surveillance

Nonconforming situations

References

INDEX

A

Aerodynamic efficiency (basic stage), 192
Aerodynamic losses, 44–45
Ainley and Mathieson method, 204–205
Ainley method, 225–226, 232, 234
Alternate energy, xiii
Analysis of losses (magnitude), 250–255
Annulus losses, 244
ASME PTC-6 test (performance testing), 292, 299–300, 302–304, 321–322:
  simplified, 321–322
Auxiliary cycle equipment, 8
Availability (unit), 142, 154–163:
  unavailability, 142;
  strategy, 157;
  cost of outage, 158–161;
  interval between inspections, 161–163
Axial clearances, 69
Axial deflection (diaphragm), 113
Axial entry (rotating blade root form), 129
Axial thrust, 47–49, 27, 114, 125

B

Base load unit, xii
Basic profile loss (blade geometry), 196–205, 262:
  Traupel method, 198–200;
  Craig and Cox method, 200–203;

Chen method, 203;
Horlock methods, 204–205;
damage, 262
Basic stage (aerodynamic) efficiency, 192
Bid specification (heat balance), 5–7
Blade bypass (steam leakage), 260–261
Blade discharge height, 68
Blade elements (analysis), 144–146
Blade extremities, 172, 175–177
Blade profile geometry (group one losses), 196–234, 262:
basic profile loss, 196–205, 262;
stage Reynolds number, 205–216;
stage discharge Mach number, 217–221;
discharge trailing edge, 221–226;
suction surface trailing edge curvature, 226–229;
incidence and inlet nose, 229–234;
damage, 262
Blade row aspect ratio, 240–243:
Craig and Cox method, 241;
Horlock methods, 242–243
Blade row steam leakage, 38–42
Boiler feed pump power, 315–316
Built-up rotor form, 123

C

Calibration (performance test instrumentation), 303–321
Calorimeter method (steam quality measurement), 319
Capacity availability, xi–xii, xv–xvi
Casing radial growth, 113
Casings (steam path), 113, 118–120:
radial growth, 113;
high pressure/high temperature, 118–119;
low pressure, 119–120
Cavity losses, 245
Chen method, 193–194, 203, 219–221, 228, 232–233

Classes (performance testing), 279–283:
manufacturer request, 280–281;
specific test code, 281–282;
code standards conducted by owner, 282;
no specific code, 283
Code tests, 281–283, 291–292
Codes (performance testing), 281–283, 291–292:
specific code, 281–282;
code standards, 282;
no specific code, 283;
other, 291–292
Combined cycle, xii–xiii
Combined rotor leakage, 23–26
Combined uncertainties (instrumentation), 303–304
Components (steam path), 118–136: casings, 118–120;
diaphragms, 120–122;
stationary blade rows, 120–122;
rotors, 122–127;
rotating blades, 127–131;
stage hardware, 131–135;
sealing system, 135–136;
development of new products/processes, 136–143;
final design, 143–148
Condensate extraction pump, 17, 71
Condition prior to test (performance), 298
Configurations (low pressure extraction), 91–98:
sections with symmetric extractions, 91–92;
sections with non-symmetric extractions, 91–92;
four-flow units, 92–93;
six-flow units, 94–98
Construction material, 103
Control stages, 58–64, 80
Correction factors (performance testing), 298–299
Cost of damage mechanisms, 256–263:
steam leakage bypassing blades, 260–261;
steam leakage at shaft-end glands, 261;
loss due to surface deposits, 261;
profile damage losses, 262;
discharge area deformation losses, 263

Cost of outage, 158–161:
replacement energy cost, 158–159;
maintenance costs, 159–160;
reserve capacity cost, 160;
changes in operating cost, 160–161
Cost of performance testing, 277–279
Coverband losses, 245–246
Coverbands, 131, 133–135, 245–246
Craig and Cox method, 195, 200–203, 211–212, 219, 223–224, 228–229, 231–232, 235–236, 238–239, 241
Critical speeds (rotor), 112
Cycle performance, 7–8
Cylindrical design (rotating blade), 127–128

D

Damage mechanisms cost, 256–263:
example, 259–260;
steam leakage bypassing blades, 260–261;
steam leakage at shaft-end glands, 261;
loss due to surface deposits, 261;
profile damage losses, 262;
discharge area deformation losses, 263
Defining (turbine design), 3–4
Definition/elements of performance, 151–154
Deflection under load (diaphragm), 113
Deformation losses (discharge area), 263
Degree of reaction, 36–38
Derated unit, 141–142
Design (mechanical), 101–148:
design process, 102–115;
section arrangement, 115–117;
major components, 118–136;
casings, 118–120;
diaphragms, 120–122;
stationary blade rows, 120–122;
rotors, 122–127;

rotating blades, 127–131;
stage hardware, 131–135;
sealing system, 135–136;
development of new products/processes, 136–143;
final design, 143–148
Design (thermodynamic), 1–99:
design process, 3–26;
section and stage energy, 26–27;
philosophy and specification, 28–38;
state-line efficiency, 38–42;
velocity/vector diagrams, 42–57;
steam path sizing and arrangement, 58–64;
radial pressure gradient, 65–67;
stage construction details, 67–69;
feed water heating trains, 69–77;
flow splitting and steam extraction, 77–98;
references, 99
Design philosophy and specification (thermodynamics), 28–38:
energy release in stage, 28–31;
stage selection, 31–38;
sensitivity of stage efficiency to ρ, 33–38
Design process (thermodynamics), 3–26:
defining the turbine, 3–4;
heat balance, 5–26
Design review (steam path), 146
Development (new products/processes), 136–143:
innovation, 137–141;
operational problem, 141–143
Diaphragm axial deflection under load, 113
Diaphragm key loads, 114
Diaphragm support system, 114
Diaphragms (steam path), 113–114, 120–122:
axial deflection, 113;
key loads, 114;
support system, 114;
stationary vane design, 121–122;
nozzle plate, 122
Differential expansion, 113
Differential pressure, 306–308

Diffuser location, 186
Dimensional drawings (manufacturing), 147
Disc friction, 237–239:
  Stodola method, 237–238;
  Ribrary method, 238;
  Craig and Cox method, 238–239
Discharge area deformation losses, 263
Discharge trailing edge, 221–229:
  Craig and Cox method, 223–224;
  Horlock methods, 224–226;
  curvature, 226–229
Double-flow sections (extraction), 79–80, 82–92, 116:
  high pressure, 79–80;
  control stages, 80;
  reheat sections, 82–88
Double-flow, high pressure sections, 79–80:
  control stages, 80
Drain flow and heat balance measurement (steam quality), 320

E

Economics/test costs (performance), 277–279
Efficiency (equipment heat balance), 7–8
Efficiency and design process (state-line), 169–171:
  stage leakage loss, 170;
  stage moisture loss, 170;
  stage leaving loss, 170;
  stage geometry, 171
Electric power, xi–xvi, 15–16, 271–272, 275–276, 311–315:
  demand, xi–xvi;
  output, 15–16, 271–272, 275–276, 311–315
Electrical power output, xi–xvi, 15–16, 271–272, 275–276, 311–315:
  testing, 271–272, 275–276, 311–315
Energy conversion process, 2
Energy ranges (sections), 26
Energy release in stage, 28–31

Enthalpy drop test (EDT), 286–291, 299, 303:
conduct of test, 286–289;
using results, 290–291
Equipment arrangement (heat balance), 8–21
Exciter power measurement, 314
Exhaust hood design, 186
Exhaust loss, 184–192:
hood, 186, 189;
hood design, 186;
turn-up loss, 189–190;
total loss, 190–192
Expansion efficiency, xix, 263–268:
example, 266–268
Expansion line end point, 12
Extraction (high pressure/reheat double-flow sections), 82–88
Extraction of partially expanded steam (high pressure sections), 81–82: feed heating extraction, 82

F

Feed heating extraction, 75–77, 82:
requirements, 75–77
Feed water heating trains, 8, 18–20, 69–78, 82:
basic considerations, 70–71;
arrangement, 71–77;
thermal conditions, 72–74;
thermodynamic requirements, 74;
extraction requirements, 75–77
Final design (steam path), 143–148:
stage mechanical analysis, 144–146;
design review, 146;
manufacturing specification, 147–148
Finite element analysis, 144
Flare, wall end, and stage transition losses, 243–246:
annulus losses, 244;
cavity losses, 245;
lap losses, 245–246

Flow measuring devices, 305–311:
primary flow, 305–310;
secondary flow, 310–311
Flow quantities measurement
(test instrumentation/calibration), 304–311:
devices, 305–311;
primary flow, 305–310;
secondary flow, 310–311
Flow reversal, 81–82
Flow splitting and steam extraction, 77–98:
double-flow, high pressure sections, 79;
double-flow control stages, high pressure sections, 80;
extraction of partially expanded steam,
high pressure sections, 81–82;
extraction from high pressure/reheat double-flow sections, 82–88;
low pressure extractions from multi-flow sections, 88–91; possible
low-pressure extraction configurations, 91–98
Force diagram, 47
Forster method, 212–216
Four-flow units (extraction), 92–93
Friction losses, 44–45
Fuel costs, 278–279
Functional tests, 104, 148:
design process, 104;
manufacturing, 148

G

Gas topping unit, xii
Generator electrical power output, xi–xvi, 16–17, 271–272, 275–276,
311–315:
testing, 271–272, 275–276, 311–315
Guarantee extent (performance testing), 273–274, 277, 296

H

Heat balance (turbine design), 5–26, 58, 320:
- bid specification, 5–7;
- efficiency, 7–8;
- steam conditions, 7;
- equipment efficiency, 7–8;
- equipment arrangement, 8–21;
- valve-stem leakage, 22–23;
- shaft-end packing leakage, 23;
- combined rotor leakage, 23–26;
- steam quality measurement, 320

Heat engine concept, 2

Heat exchange vessels, 20

Heat rate, 163–167, 263–268:
- state-line efficiency influence, 263–268;
- example, 266–268

Heater above reheat point, 81

Heater extraction points, 6

Heating trains (feed water), 6, 18–20, 69–77:
- basic considerations, 70–71;
- arrangement, 71–77;
- thermal conditions, 72–74;
- thermodynamic requirements, 74;
- extraction requirements, 75–77

High pressure sections, 79–88:
- double flow, 79–80;
- double flow control, 80;
- extraction of partially expanded steam, 81–82;
- double-flow extraction, 82–88

High pressure/high temperature casing, 118–119

High pressure/reheat double-flow sections (extraction), 82–88

High reaction (two-dimensional considerations), 49–50, 52–55:
- example, 52–55

High temperature/high pressure casing, 118–119

High-pressure sections, 12–14, 22, 26–27, 79–88, 117–119:
- reheat double flow, 82–88;
- high temperature, 118–119

Hood exhaust loss, 186, 189
Horlock methods (described), 204–205, 224–226, 232, 234, 242–243:
Ainley and Mathieson method, 204–205;
Markov method, 225;
Ainley method, 225–226, 232, 234;
Soderberg method, 232, 234

I–J

Impulse stage design, 28, 34–36, 43–52, 55, 67:
comparison with reaction stage, 35;
sensitivity, 36;
pure impulse, 43–52, 55;
two-dimensional considerations, 43–52, 55;
example, 50–52, 55
Incidence and inlet nose, 229–234:
Craig and Cox method, 231–232;
Chen method, 232–233;
Horlock methods, 232, 234
Information from thermal design (mechanical design process), 104–110:
steam mass flow rate, 104–105;
steam conditions, 105–110
Inlet energy, 7
Inlet nose (incidence), 229–234:
Craig and Cox method, 231–232;
Chen method, 232–233;
Horlock methods, 232, 234
Innovation (new products/processes), 137–141
Inspection interval, 161–163: factors, 162–163
Instrument requirements (performance testing), 302
Instrument schedule (performance testing), 295
Instrument type/location/connections (performance testing), 297
Instrumentation and calibration (performance testing), 302–321:
instrument requirements, 302;
combined uncertainties, 303–304;
flow quantities measurement, 304–311;
power output measurement, 311–315;

boiler feed pump power, 315–316;
thermal energy measurement, 316–319;
steam quality measurement, 319–320;
time measurement, 320;
unit speed measurement, 321
Isolation (performance test cycle), 300–301

K

Key loads (diaphragm), 114
Kinetic energy, 2, 48–49, 165, 168–169, 184–188

L

Labor cost, 159–160:
personnel, 159;
cleaning parts, 159;
specialized maintenance, 159–160
Lacing wire or tie losses, 235–236:
Craig and Cox method, 235–236
Lap efficiency/optimization, 58–64
Lap losses, 245–246
Laser probe traverse method (steam quality measurement), 320
Leakage (heat balance), 22–26:
valve stem, 22–23;
shaft end, 23;
rotor, 23–26
Leakage steam effect (state-line efficiency), 178–181
Load values (performance testing), 296
Loss analysis (magnitude), 149–151, 250–255, 271–272
Low pressure casing, 119–120

Low pressure extraction configurations, 91–98:
symmetric extractions, 91–92;
non-symmetric extractions, 91–92;
four-flow units, 92–93;
six-flow units, 94–98
Low pressure extractions from multi-flow sections, 88–91
Low-pressure sections, 16–18, 23, 76–77, 88–98, 115, 117, 119–120, 185:
multi-flow, 88–91;
extraction configurations, 91–98

M

Mach number (stage discharge), 217–221:
Craig and Cox method, 219;
Chen method, 219–221
Maintenance costs, 159–160:
labor cost, 159;
cleaning parts, 159;
specialized maintenance, 159–160
Maintenance, x, xiv–xvi, 159–160:
costs, 159–160
Manpower responsibility (performance testing), 297
Manufacturer request (performance testing), 280–281
Manufacturing design, 143–148:
mechanical analysis, 144–146;
design review, 146;
manufacturing specification, 147–148
Manufacturing options (rotor), 127
Manufacturing specification (steam path), 143–148:
final design, 143–148;
dimensional drawings, 147;
material specification, 147;
method of material manufacture, 147;
special processes, 147;
functional tests, 148;
non-destructive tests, 148

Manufacturing, x, xviii–xix, 127, 143–148:
rotor options, 127;
steam path design, 143–148;
steam path specification, 147–148
Margins applied to results (performance testing), 299–300
Markov method, 225
Material manufacturing, 147:
method, 147;
specification, 147
Mechanical analysis (stage), 141, 144–146
Mechanical considerations (design process), x, xx, 110–115, 141, 144–146:
rotor critical speeds, 112;
shaft deflected form, 112–113;
diaphragm axial deflection under load, 113;
short circuit torque, 113;
differential expansion, 113;
rotor and casing radial growth, 113;
axial thrust, 114;
diaphragm key loads, 114;
diaphragm support system, 114;
mechanical analysis, 141, 144–146
Mechanical design process, x, xx, 102–115, 141, 144–146:
tolerance of manufacture, 103;
surface finish requirements, 103;
material of construction, 103;
special processes, 104;
special functional tests, 104;
non-destructive tests, 104;
information required from thermal design, 104–110;
mechanical considerations, 110–115;
mechanical analysis, 141, 144–146
Mechanical drive power measurement, 314–315
Moisture content/damage, 27
Moisture loss, 27, 181–184: damage, 27;
effect on state-line efficiency, 182–184
Mollier diagram, 26–27, 37, 41, 106, 164–165, 288
Monoblock rotor form, 123
Multi-flow sections, 88–92:
low pressure extractions, 88–91

N

Negative root reaction, 248–250
New products/processes (steam path mechanical design), 136–143:
innovation, 137–141;
operational problem, 137–143
Non-conforming conditions, 154–158
Non-destructive tests, 104, 148, 157:
design process, 104; manufacturing, 148
Non-symmetric extraction, 91–92
Nozzle plate, 122
Nozzle repair, 97
Nuclear applications, 79, 82–83, 311

O

Objectives (performance testing), 274–277, 284–285, 296
Operating cost changes, 160–161
Operating period, x, xiv–xv
Operational problem (new products/processes), 137–143:
innovation, 137–141;
unit derated, 141–142;
unit no longer available, 142
Outage cost, 158–161:
replacement energy cost, 158–159;
maintenance costs, 159–160;
reserve capacity cost, 160;
changes in operating cost, 160–161
Output loss (damage), 256–263:
example, 259–260;
steam leakage, 260–261;
surface deposits, 261;
profile, 262;
discharge area deformation, 263
Output test, 271–272, 311–315

P–Q

Percentage reaction stage design, 49–50, 52–55:
example, 52–55
Performance (steam path), xx, 149–270:
performance definition/elements, 151–154;
factors, 152–154;
steam path availability, 154–163;
state–line efficiency, 163–192;
stage efficiency considerations, 193–195;
blade profile geometry group one losses, 196–234;
tie or lacing wire losses, 235–236;
stage geometric factors group two losses, 236–250;
analysis of magnitude of losses, 250–255;
cost of damage mechanisms, 256–263;
influence of state-line efficiency on heat rate, 263–268;
references, 269–270.
See also Steam power cycle (performance testing).
Performance acceptance test, 272–277
Performance definition/elements (steam path), 151–154
Performance factors, 152–154
Performance test instrumentation, 302–321:
instrument requirements, 302;
combined uncertainties, 303–304;
flow quantities measurement, 304–311;
power output measurement, 311–315;
boiler feed pump power, 315–316;
thermal energy measurement, 316–319;
steam quality measurement, 319–320;
time measurement, 320;
unit speed measurement, 321
Performance testing (steam power cycle), xx, 271–323:
purpose, 272–277;
test costs and economics, 277–279;
classes, 279–283;
test objectives and requirements, 284–285;
enthalpy drop test, 286–291;
other test codes, 291–292;

test schedule and advanced planning, 293–300;
test cycle isolation, 300–301;
test instrumentation and calibration, 302–321;
simplified ASME PTC-6 test, 321–322;
references, 323.
See also Steam path performance.
Performance testing classes, 279–283:
manufacturer request, 280–281;
specific test code, 281–282;
code standards conducted by owner, 282;
no specific code, 283
Performance testing codes, 281–283, 291–292:
specific code, 281–282;
code standards, 282;
no specific code, 283;
other, 291–292
Performance testing purpose, 272–277:
performance acceptance, 272–277;
predicted/guaranteed performance, 273–274, 277;
prototype/new design, 273–274;
parameters, 275–276;
steam quantity, 275;
thermal energy, 275;
power generated, 275–276
Power factor measurement, 314
Power generated, xi–xvi, 15–16, 271–272, 275–276, 311–315:
output measurement, 311–315
Power output measurement, 311–315:
generator electrical power, 312–314;
power factor, 314;
exciter power, 314;
mechanical drive power, 314–315
Pressure balance holes, 247–248
Pressure balance piston, 172, 178
Pressure differential, 18, 24–25
Pressure drop, 18, 24–25, 109
Pressure energy, 165, 184
Pressure measurement, 316–318
Primary flow measuring, 305–310
Profile damage losses, 262

Profile loss (blade geometry), 196–205, 262:
Traupel method, 198–200;
Craig and Cox method, 200–203;
Chen method, 203;
Horlock methods, 204–205;
damage, 262
Prototype/new design (testing), 273–274
Purchase agreement, xiii
Pure impulse stage, 43–52, 55:
two-dimensional considerations, 43–52, 55;
example, 50–52, 55
Purpose (performance testing), 272–277:
performance acceptance, 272–277;
predicted/guaranteed performance, 273–274, 277;
prototype/new design, 273–274;
parameters, 275–276;
steam quantity, 275;
thermal energy, 275;
power generated, 275–276

R

Radial clearances, 69
Radial entry (rotating blade root form), 130
Radial flow, 55–57
Radial growth (rotor and casing), 113
Radial pressure gradient, 65–67
Reaction stage design, 28, 33, 35–38, 67:
comparison with impulse stage, 35;
degree of reaction, 36–38
Regenerative feed water heating, 77–78
Reheat effect, 40–41
Reheat point, 81
Reheat sections, 82–88, 116:
double-flow (extraction), 82–88
Renewable energy sources, xiii
Replacement energy cost, 158–159

Reserve capacity cost, 160
Reverse flow, 81–82
Reynolds number (stage), 205–216:
  Traupel method, 208–211;
  Craig and Cox method, 211–212;
  Forster method, 212–216
Ribrary method, 238
Risk/risk analysis, 139–140, 150–151:
  development project, 139–140
Root form selection (rotating blade), 128–131:
  axial entry, 129;
  radial entry, 130;
  tangential entry, 130–131
Rotating blade row, xix, 38–42, 45–46, 56, 180–181
Rotating blades (steam path), xix, 38–42, 45–46, 56, 107–108, 122–123, 127–131, 157, 180–181:
  blade row, xix, 38–42, 45–46, 56, 181;
  cylindrical design, 127–128;
  vortex/twisted design, 128;
  vortex design approximate, 128;
  selection of root form, 128–131
Rotor and casing radial growth, 113
Rotor configuration, 123–127:
  technical considerations, 126–127
Rotor critical speeds, 112
Rotor form design, 123, 125–126
Rotor interchangeability, 89–90, 95–96
Rotor leakage, 23–26
Rotor materials, 126
Rotor radial growth, 113
Rotors (steam path), 23–26, 89–90, 95–96, 112–113, 122–127:
  leakage, 23–26;
  interchangeability, 89–90, 95–96;
  critical speeds, 112;
  casing radial growth, 113;
  configuration, 123–127;
  form design, 123, 125–126;
  materials, 126;
  technical considerations, 126–127

Row aspect ratio, 240–243:
Craig and Cox method, 241;
Horlock methods, 242–243

S

Schedule (performance testing), 297
Seal/sealing, 96, 121, 135–136, 173–175
Sealing system (steam path), 96, 121, 135–136, 173–175:
blade row, 96;
seal system design, 135–136
Secondary flow measuring, 310–311
Section arrangement (steam path), 115–117
Section energy design, 26–27
Sensitivity (stage efficiency), 33–38:
reaction design, 33;
impulse design, 34;
reaction/impulse stage comparison, 35;
impulse stage, 36;
degree of reaction, 36–38
Sensitivity analysis, 6
Shaft deflected form, 112–113
Shaft-end glands (leakage), 261
Shaft-end leakage, 23, 172, 177, 261:
packing, 23;
glands, 261
Shaft-end packing (leakage), 23
Short circuit torque, 113
Simplified ASME PTC-6 test, 321–322
Six-flow units (extraction), 94–98
Size limitations (rotor), 126
Soderberg method, 232, 234
Special functional tests (design process), 104
Special processes, 104, 147:
design process, 104;
manufacturing, 147
Specialized maintenance cost, 159–160

Stage construction details, 67–69:
stage diameters, 68;
blade discharge heights, 68;
stage laps, 68;
radial clearances, 69;
axial clearances, 69
Stage diameter, 68, 108
Stage discharge Mach number, 217–221:
Craig and Cox method, 219;
Chen method, 219–221
Stage efficiency considerations, 31–38, 193–195:
optimization, 31–33;
sensitivity, 33–38
Stage efficiency sensitivity, 33–38:
reaction design, 33;
impulse design, 34;
reaction/impulse stage comparison, 35;
impulse stage, 36;
degree of reaction, 36–38
Stage energy design, 26–27
Stage energy release, 28–31
Stage geometric factors (group two losses), 171, 193–195, 236–250:
disc friction, 237–239;
row aspect ratio, 240–243;
wall end, flare and stage transition losses, 243–246;
pressure balance holes, 247–248;
negative root reaction, 248–250
Stage geometry (efficiency), 171, 193–195, 236–250
Stage hardware (steam path), 131–135
Stage laps, 58–64, 68, 245–246:
optimization, 58–64;
losses, 245–246
Stage leakage losses, 170–178:
blade extremities, 172, 175–177;
shaft end leakage, 172, 177;
pressure balance piston, 172, 178;
valve stem leakage, 172, 178

Stage leaving loss, 170, 184–192:
velocity loss, 186–188;
hood loss, 189;
turn-up loss, 189–190;
total exhaust loss, 190–192
Stage losses, 38–42, 170–178, 181–192:
moisture, 170, 181–184;
leaving, 170, 184–192;
leakage, 170–178
Stage mechanical analysis (steam path), 144–146
Stage moisture loss, 170, 181–184
Stage Reynolds number, 205–216:
Traupel method, 208–211;
Craig and Cox method, 211–212;
Forster method, 212–216
Stage selection, 31–38:
optimization, 31–33;
sensitivity of stage efficiency to ρ, 33–38
Stage transition, wall end, and flare losses, 243–246:
annulus losses, 244;
cavity losses, 245;
lap losses, 245–246
State-line efficiency, 38–42, 163–192, 263–268:
total-to-total efficiency, 167;
total-to-static energies, 168–169;
efficiency and the design process, 169–171;
stage leakage losses, 171–178;
leakage steam effect, 178–181;
stage moisture loss, 181–184;
stage leaving loss, 184–192;
basic stage aerodynamic efficiency, 192;
influence on heat rate, 268–268;
example, 266–268
Stationary blade row (steam path), 43–44, 46, 56, 107, 120–122, 179–181:
stationary vane design, 121–122;
nozzle plate, 122
Stationary blades, 43–44, 46, 56, 107, 120–122, 179–181:
blade row, 43–44, 46, 56, 107, 120–122, 179–181

Stationary vane design, 121–122
Steam conditions (performance testing), 296:
  allowable deviations, 296
Steam conditions, 5, 7–8, 10–12, 105–110, 126, 296:
  heat balance, 7;
  surrounding rotor, 126;
  performance testing, 296
Steam extraction, 75–98:
  considerations, 75–91, 96–98;
  flow splitting, 77–98;
  double-flow, high pressure sections, 79;
  double flow control stages, high pressure sections, 80;
  partially expanded steam, high pressure sections, 81–82;
  high pressure/reheat double-flow sections, 82–88;
  low pressure, multi-flow sections, 88–91;
  possible low-pressure configurations, 91–98
Steam leakage, 38–42, 135–136, 250–255, 260–261:
  blade row, 38–42;
  loss analysis, 250–255;
  bypassing blades, 260–261;
  shaft-end glands, 261
Steam mass flow rate, 104–105
Steam packing exhauster, 20
Steam path availability, 154–163:
  strategy, 157;
  cost of outage, 158–161;
  interval between inspections, 161–163
Steam path losses (analysis), 250–255
Steam path mechanical design, 101–148:
  design process, 102–115;
  section arrangement, 115–117;
  major components, 118–136;
  casings, 118–120;
  diaphragms, 120–122;
  stationary blade rows, 120–122;
  rotors, 122–127;
  rotating blades, 127–131;
  stage hardware, 131–135;
  sealing system, 135–136;

development of new products/processes, 136–143;
final design, 143–148;
review, 146
Steam path performance, xix–xx, 149–270:
performance definition/elements, 151–154;
steam path availability, 154–163;
state-line efficiency, 163–192;
stage efficiency considerations, 193–195;
blade profile geometry group one losses, 196–234;
tie or lacing wire losses, 235–236;
stage geometric factors group two losses, 236–250;
analysis of magnitude of losses, 250–255;
cost of damage mechanisms, 256–263;
influence of state-line efficiency on heat rate, 263–268; references, 269–270.
See also Steam power cycle (performance testing).
Steam path sizing/arrangement, 58–64:
solution of vane heights, 58–64
Steam path, xix–xx, 8, 10–12, 58–64, 101–270:
performance, xix–xx, 149–270;
sizing/arrangement, 58–64;
mechanical design, 101–148;
availability, 154–163;
loss analysis, 250–255
Steam power cycle (performance testing), xx, 271–323:
purpose, 272–277;
test costs and economics, 277–279;
classes, 279–283;
test objectives and requirements, 284–285;
enthalpy drop test, 286–291;
other test codes, 291–292;
test schedule and advanced planning, 293–300;
test cycle isolation, 300–301;
test instrumentation and calibration, 302–321;
simplified ASME PTC-6 test, 321–322;
references, 323.
See also Steam path performance.

Steam quality measurement, 319–320:
calorimeter method, 319;
drain flow and heat balance, 320;
tracer injection techniques, 320;
laser probe traverse method, 320
Steam quantity (performance testing), 275
Steam seal regulator, 20–21
Steam turbine concept, 2–3, 8
Stodola method, 237–238
Stream lines, 42
Suction surface trailing edge curvature, 226–229:
Chen method, 228;
Craig and Cox method, 228–229
Supplementary tests (performance), 297
Support system (diaphragm), 114
Surface deposits loss, 261
Surface finish requirements, 103
Symmetric/non-symmetric extraction, 83–84, 86–88, 91–92, 98

T

Tangential entry (rotating blade root form), 130–131
Technical considerations (rotor configuration), 126–127:
size limitations, 126;
steam condition surrounding rotor, 126;
rotor materials, 126;
manufacturing options, 127
Temperature measurement, 318–319
Terminal temperature difference, 18
Test classes (performance), 279–283:
manufacturer request, 280–281;
specific test code, 281–282;
code standards conducted by owner, 282;
no specific code, 283
Test codes (performance), 271, 281–283, 291–292:
specific code, 281–282;

code standards, 282;
no specific code, 283;
other, 291–292
Test conference (performance), 293–300:
definition of objectives, 296;
load values for test, 296;
extent of guarantee, 296;
steam conditions for test, 296;
allowable deviations from steam conditions, 296;
responsibility for manpower, 297;
instrument type/location/connections, 297;
supplementary tests, 297;
schedule test date, 297;
condition prior to test, 298;
proposed test run, 298;
test run duration and frequency of readings, 298;
calculation of results and application of correction factors, 298–299;
margins applied to results, 299–300
Test costs and economics (performance), 277–279
Test cycle isolation (performance), 300–301
Test instrument requirements, 302
Test instrumentation and calibration (performance), 302–321:
instrument requirements, 302;
combined uncertainties, 303–304;
flow quantities measurement, 304–311;
power output measurement, 311–315;
boiler feed pump power, 315–316;
thermal energy measurement, 316–319;
steam quality measurement, 319–320;
time measurement, 320;
unit speed measurement, 321
Test objectives/requirements (performance), 284–285
Test results calculation (performance), 298–299:
readings, 298
Test run (performance), 298:
proposed, 298;
duration, 298
Test schedule/planning (performance), 293–300:
test conference, 295–300;
test run proposed, 298

Thermal analysis, 5, 102–110, 141:
information, 104–110;
steam mass flow rate, 104–105;
steam conditions, 105–110
Thermal design process, 3–26, 102–110, 141:
defining the turbine, 3–4;
thermal conditions, 5;
basic heat balance, 5–26;
thermal analysis, 5, 102–110, 141
Thermal energy (performance testing), 275
Thermal energy measurement, 316–319:
pressure measurement, 316–318;
temperature measurement, 318–319
Thermal energy, 275, 316–319:
available/supplied, 275;
measurement, 316–319
Thermal gradient, 18
Thermodynamic design, xviii–xx, 2, 5, 42, 1–99:
principles/theory, 2;
thermal design process, 3–26;
section and stage energy, 26–27;
design philosophy and specification, 28–38;
state-line efficiency, 38–42;
thermal analysis, 42;
velocity/vector diagrams, 42–57;
steam path sizing and arrangement, 58–64;
radial pressure gradient, 65–67;
stage construction details, 67–69;
feed water heating trains, 69–77;
flow splitting and steam extraction, 77–98;
references, 99
Three-dimensional considerations, 55–57
Thrust (force), 27, 47–49, 114, 125
Tie or lacing wire losses, 235–236:
Craig and Cox method, 235–236
Tie wires, 109–110, 131–133, 235–236:
losses, 235–236
Time measurement, 320
Tolerance (manufacturing), 103
Total exhaust loss, 190–192

Total-to-static energies, 168–169
Total-to-total efficiency, 167
Tracer injection techniques (steam quality measurement), 320
Trailing edge (discharge), 221–229:
 Craig and Cox method, 223–224;
 Horlock methods, 224–226;
 curvature, 226–229
Trailing edge curvature (suction surface), 226–229:
 Chen method, 228;
 Craig and Cox method, 228–229
Traupel method, 198–200, 208–211
Turbine definition, 3–4
Turn-up loss, 189–190
Two-dimensional considerations, 43–55:
 pure impulse, 43–52;
 impulse example, 50–52;
 with high reaction, 49–50, 52–55;
 high reaction example, 52–55
Two-flow section, 117

U

Uncertainties (test instrumentation/calibration), 303–304
Unit availability, 142, 154–163:
 unavailability, 142;
 strategy, 157;
 cost of outage, 158–161;
 interval between inspections, 161–163
Unit derated, 141–142
Unit maintainability, 155–156:
 damage determination, 155;
 accessibility, 155;
 component interchangeability, 155–156
Unit speed measurement, 321
Upgrading, xii
Used energy end point, 12

V

Valve-stem leakage, 22–23, 172, 178
Vane design, 121–122
Vane heights, 58–64
Vane profiles, 109–110
Vector diagrams.
See Velocity/vector diagrams.
Velocity energy.
See Kinetic energy.
Velocity loss, 186–188
Velocity/vector diagrams, 42–57:
two-dimensional considerations (pure impulse), 43–52;
two-dimensional considerations (with high reaction), 49–50, 52–55;
three-dimensional considerations, 55–57
Vortex design approximate (rotating blade), 128
Vortex/twisted design (rotating blade), 128

W–Z

Wall end losses, 243–246:
annulus losses, 244;
cavity losses, 245;
lap losses, 245–246
Weld rebuild (nozzle), 97
Welded rotor form, 123